STANDARDS FOR PREPARING TEACHERS OF MATHEMATICS

AMTE

ASSOCIATION OF MATHEMATICS TEACHER EDUCATORS

STANDARDS FOR PREPARING TEACHERS OF MATHEMATICS WRITING TEAM

Nadine Bezuk, San Diego State University, Chair

Jennifer M. Bay-Williams, University of Louisville, Leadership Team Member

Douglas H. Clements, University of Denver, Leadership Team Member

W. Gary Martin, Auburn University, Leadership Team Member

Julia Aguirre, University of Washington Tacoma

Timothy Boerst, University of Michigan

Elizabeth A. Burroughs, Montana State University

Ed Dickey, University of South Carolina

Rochelle Gutiérrez, University of Illinois at Urbana-Champaign

Elizabeth Hughes, University of Northern Iowa

DeAnn Huinker, University of Wisconsin — Milwaukee

Karen Karp, Johns Hopkins University

W. James Lewis*, University of Nebraska-Lincoln

Travis A. Olson, University of Nevada, Las Vegas

Randolph A. Philipp, San Diego State University

Nicole Rigelman, Portland State University

Marilyn E. Strutchens, Auburn University

Christine D. Thomas, Georgia State University

Dorothy Y. White, University of Georgia

* W. James Lewis participated as a member of the Writing Team while serving at the National Science Foundation. Any opinion, findings, and conclusions or recommendations expressed in this material are those of the authors and do not necessarily reflect the views of the National Science Foundation.

Acknowledgements

Thanks to the following people for their work in the design, production, and editing of this document:

Joe Champion, Boise State University; Website Director, Association of Mathematics Teacher Educators

Timothy Hendrix, Meredith College; Executive Director, Association of Mathematics Teacher Educators

Tony Nguyen, Meredith College; Graphic Designer and Webmaster, Association of Mathematics Teacher Educators

Bonnie Schappelle, San Diego State University; Research Assistant & Copy-Editor

The work of preparing these standards was supported by a grant from the Brookhill Institute of Mathematics.

Library of Congress Cataloging-in-Publication Data

A CIP record for this book is available from the Library of Congress
http://www.loc.gov

ISBN: 978-1-64113-999-1 (Paperback)
978-1-64802-000-1 (Hardcover)
978-1-64113-998-4 (E-Book)

Copying and Reprinting

Printed and bound by CPI Group (UK) Ltd, Croydon, CR0 4YY

Inquiries and requests can also be made by e-mail to amte-support@amte.net, or by mail to the Association of Mathematics Teacher Educators, c/o Meredith College, 3800 Hillsborough Street, Raleigh, NC 27607.

CONTENTS

TABLES

FIGURES

VIGNETTES

FOREWORD

The Association of Mathematics Teacher Educators (AMTE), uniquely positioned as the lead organization for mathematics teacher education in the United States, puts forth these standards as a national guide for the preparation of prekindergarten through Grade 12 (Pre-K–12) teachers of mathematics. The mission of AMTE is to promote the improvement of mathematics teacher education Pre-K–12 with stated goals focused on effective mathematics teacher education programs and effective policies and practices related to mathematics teacher education at all levels. Over the 25-year history of AMTE, the organization has produced three standards documents: *Principles to Guide the Design and Implementation of Doctoral Programs in Mathematics Education* (2002), designed for institutions of higher education to guide review, revision, or creation of doctoral programs in mathematics education; *Standards for Elementary Mathematics Specialists* (first published 2009, updated 2013), designed to define and advocate for effective preparation of mathematics specialists; and, now, the *Standards for the Preparation of Teachers of Mathematics* (2017), created to address issues and challenges facing teacher preparation and articulate a national and comprehensive vision for the initial preparation of teachers of mathematics in Pre-K–12.

The AMTE Board of Directors' decision to initiate the development of standards was ignited by Dr. Nadine Bezuk of San Diego State University during her delivery of the Judith Jacobs Lecture (JJL) at the 2015 AMTE annual conference. The JJL was established in 2003 to honor Dr. Judith E. Jacobs, one of the founding members of AMTE. Since that time a renowned mathematics educator has been selected each year to deliver the JJL at the annual conference. The focus of Dr. Bezuk's lecture was the need for AMTE to provide leadership to the field by developing standards for preparing teachers to teach mathematics. She argued, "While there exist a number of documents that address various aspects of the initial preparation of mathematics teachers, there is no single, definitive document addressing the initial preparation of mathematics teachers more globally." Further, she noted an absence, within the mathematics teacher education community, of a shared vision of what preparing teachers entails and a limited understanding of the degree to which any such vision may be shared with other stakeholders involved in the initial preparation of mathematics teachers. Unequivocally, the AMTE Board of Directors was challenged by the lecture to develop and disseminate national standards, and as a result, the AMTE Board of Directors made the decision to develop and disseminate Pre-K–12 standards for mathematics teacher preparation.

In March 2015, the AMTE Board of Directors established a leadership team to develop the standards. The team includes Douglas H. Clements of the University of Denver as lead developer for the early childhood grades; Nadine Bezuk as chair and lead developer for upper elementary grades; Jennifer Bay-Williams of the University of Louisville as lead developer for the middle level; and W. Gary Martin of Auburn University for high school.

The leadership team began by establishing criteria for the expertise required to write the standards. The leadership team sought, as a collective body of knowledge across the members of the writing team, expertise that included experiences in the preparation of mathematics teachers at the early childhood, upper elementary, middle school, or high school level; the teaching of mathematics methods courses; the teaching of mathematics coursework and development of mathematical knowledge for teaching; the supervision of clinical placements and field experiences; the responsibility for recruiting and retaining students in mathematics teacher education programs; pedagogy to support emergent multilingual learners; pedagogy to support special needs students; advocacy for equity in mathematics teaching and learning; and the administration of mathematics teacher education policy and change agency. With these criteria in mind, the leadership team identified and invited mathematics educators and mathematicians to serve as members of the writing team. The members of the writing team are listed in the document. The AMTE Board of Directors extends sincere appreciation and gratitude to the members of the writing team. These individuals were dedicated and committed to the significance of this work and the potential of the influence of these standards for improving mathematics teaching.

Another noteworthy aspect of the process was the attention given and time invested in a coherent and comprehensive review of drafts. In the initial planning phase, the leadership team established and implemented a review process to ensure that a broad range of stakeholders would have opportunities to provide critical feedback throughout the development stages. As such, a group of mathematics educators was identified and invited to serve on a review team to facilitate or assist in development of the document while the writing group engaged in writing and refining chapters. The AMTE Board of Directors especially thanks the review team for providing feedback on the drafts of the chapters on an ongoing basis. The following were members of the review team:

- Robert Q. Berry, III, University of Virginia; President-elect, National Council of Teachers of Mathematics
- Francis (Skip) Fennell, McDaniel College; Past President, Association of Mathematics Teacher Educators; Past President, National Council of Teachers of Mathematics; Vice Chair, Council for the Accreditation of Educator Preparation Board of Directors
- Matt Larson, Lincoln (Nebraska) Public Schools and President, National Council of Teachers of Mathematics
- Dale Oliver, Humboldt State University (California)
- Margaret (Peg) Smith, University of Pittsburgh

Additionally, announcements and special requests for review of the draft standards went out to AMTE members, other key stakeholders, and professional organizations representing mathematics education and the mathematical sciences. The AMTE Board of Directors sincerely appreciates all individuals who provided feedback on the draft document and professional organizations for assembling teams to provide feedback on their behalf. We received feedback from the following organizations:

- American Mathematical Association of Two-Year Colleges (AMATYC)
- American Mathematical Society (AMS)
- American Statistical Association (ASA)
- Association for Middle Level Education (AMLE)
- Association for Women in Mathematics (AWM)
- Association of Mathematics Teacher Educators of Alabama (AMTEA)
- Association of State Supervisors of Mathematics (ASSM)
- Benjamin Banneker Association (BBA)
- Council of Chief State School Officers (CCSSO)
- Education Development Center (EDC)
- Hoosier Association of Mathematics Teacher Educators (HAMTE)
- Mathematical Association of America (MAA)
- National Association of Mathematicians (NAM)
- National Council of Supervisors of Mathematics (NCSM)
- National Council of Teachers of Mathematics (NCTM)
- Society for Industrial and Applied Mathematics (SIAM)
- TODOS: Mathematics for All

The *Standards for Preparing Teachers of Mathematics* are aspirational, advocating for practices that support candidates in becoming effective teachers of mathematics who guide student learning. These standards will guide the improvement of individual teacher preparation programs and promote national dialogue and action related to the preparation of teachers of mathematics.

Christine D. Thomas
AMTE President (2015–2017)

PREFACE

The future mathematical success of our nation's children is largely dependent on the teachers of mathematics they encounter from prekindergarten to Grade 12 (Pre-K–12). According to Tatto and Senk (2011), "If the quality of education for every child is to be improved, the education of teachers needs to be taken seriously" (p. 134). Those involved in preparing teachers of mathematics must ensure that all their candidates have the knowledge, skills, and dispositions to provide all students access to meaningful experiences with mathematics.

The Association of Mathematics Teacher Educators (AMTE) is the largest U.S. professional organization devoted to the preparation of teachers of mathematics. AMTE includes more than 1,000 members supporting preservice teacher education and professional development of teachers of mathematics at all levels from Pre-K–12. AMTE members include professors, researchers, teacher leaders, school-based and district mathematics supervisors and coordinators, policymakers, graduate students, and others. The Standards described in this document reflect AMTE's leadership in shaping the preparation of Pre-K–12 teachers of mathematics, including clearly articulated expectations for what *well-prepared beginning mathematics teachers* need to know and be able to do upon completion of a certification or licensing program and the characteristics such programs must have to support teachers' development.

Although the field continues to conduct research regarding effective practices for teacher preparation, we have a growing research base describing what teaching practices affect student learning and student experiences in mathematics classrooms. As an example, research indicates that focusing only on teachers' behaviors has a smaller effect on teachers' development and subsequently on their students' learning than does focusing on teachers' knowledge of the subject, on the curriculum, or on how students learn the subject (Carpenter, Fennema, Peterson, & Carey, 1988; Kennedy, 1998; Kwong et al., 2007; Philipp et al., 2007).

A number of recent documents address various aspects of the initial preparation of mathematics teachers.[1] Table O.1 summarizes their focus. Although all these documents inform mathematics teacher preparation, no single, comprehensive document addresses the initial preparation of mathematics teachers across Pre-K–12. AMTE's goal is for the standards in this document to provide a clear, comprehensive vision for initial preparation of teachers of mathematics. Building on the documents in Table 0.1, we, in this document's standards, elaborate what beginning teachers of mathematics must know and be able to do as well as the dispositions they must have to increase equity, access, and opportunities for the mathematical success of each student. Given the challenges that teachers of mathematics face in preparing their students for future success, mathematics teacher educators must be guided by a well-articulated vision to prepare teachers of mathematics to meet those challenges. In this document, we take up that charge.

TABLE 0.1. STANDARDS AND REPORTS RELATED TO PREPARING TEACHERS OF MATHEMATICS

Standards and Reports Specific to Mathematics Teachers

***The Mathematical Education of Teachers II* (*MET II*)**

The *MET II* (Conference Board of Mathematical Sciences [CBMS], 2012) addressed the mathematical content knowledge well-prepared beginning teachers of mathematics should know at the elementary, middle and high school levels.

---

[1] For the purposes of this document, mathematics teacher preparation includes preparation to teach statistics, following common practice. However, we recognize that statistics and statistics education, although related to mathematics and mathematics education, are distinct. This document will use the term mathematics to encompass mathematics and statistics; when the distinction between mathematics and statistics is important to emphasize, statistics will be identified separately.

National Council of Teachers of Mathematics' Council for the Association of Educator Preparation (CAEP) Standards

The *National Council of Teachers of Mathematics (NCTM) CAEP Standards* (NCTM & CAEP, 2012a, 2012b) described what effective preservice teachers of secondary mathematics should know and be able to do, informing program reviews for middle and high school mathematics programs.

Statistical Education of Teachers (SET)

SET (Franklin et al., 2015) describes the statistical content knowledge well-prepared beginning teachers of mathematics should know at the elementary, middle and high school levels.

Teacher Education and Development Study in Mathematics (TEDS-M)

In the *TEDS-M*, researchers examined and discussed findings and challenges related to the mathematics education of future primary, middle, and high school teachers (Tatto & Senk, 2011).

Standards Not Specific to Mathematics That Also Apply to Teachers of Mathematics

Association for Middle Level Education (AMLE) Middle Level Teacher Preparation Standards with Rubrics and Supporting Explanations

AMLE (2012) describes five standards, including one on content, which addresses content, standards for students, and the interdisciplinary nature of knowledge.

Council for Exceptional Children (CEC) Initial Preparation Standards

The CEC (2012) standards require that beginning professionals understand and use mathematics concepts to individualize learning for students.

Council for the Accreditation of Educator Preparation (CAEP) Accreditation Standards

The CAEP (2013) describes candidate and program expectations that define quality programs. Among these is the expectation that candidates demonstrate content and pedagogical knowledge in their content domains.

National Association for the Education of Young Children (NAEYC) Standards for Initial and Advanced Early Childhood professional Preparation Programs

The NAEYC (2010) professional standards describe the importance of knowing mathematics and teaching it in ways that promote sense making and nurture positive development.

Standards for Experienced Teachers of Mathematics

Mathematics Specialists

In the *Standards for Elementary Mathematics Specialists* (2013), AMTE outlined "particular knowledge, skills, and dispositions" needed by elementary mathematics specialists who "teach and support others who teach mathematics at the elementary level." (p. iv)

Interstate Teacher Assessment and Support Consortium (InTASC) Model Core Teaching Standards

The *InTASC Model Core Teaching Standards* (Council of Chief State School Officers [CCSSO], 2013) are used in states, school districts, professional organizations, and teacher education programs to support teachers.

National Board for Professional Teaching Standards (NBPTS) Standards

The NBPTS standards recognize accomplished teachers and included certifications for early childhood and elementary school generalists (2012a, 2012b), and middle school and high school mathematics teachers (2010).

PURPOSE

This document includes a set of comprehensive standards describing a national vision for the initial preparation of all teachers Pre-K–12 who teach mathematics. That is, in addition to early childhood and elementary school teachers who teach all disciplines, middle grade teachers, and high school mathematics teachers, these standards are also directed toward special education teachers, teachers of emergent multilingual students, and all others who have responsibility for aspects of student learning in mathematics.

These standards are intended to

- guide the improvement of individual teacher preparation programs,
- inform the accreditation process of such programs,
- influence policies related to preparation of teachers of mathematics, and
- promote national dialogue and action related to preparation of teachers of mathematics.

These standards are aspirational, advocating for mathematics teacher preparation practices that support candidates in becoming high-quality teachers who are ethical advocates for children and effectively guide student learning aligned with research and best practices, rather than describing minimum levels of competency needed by beginning teachers. The standards are intended both to build on existing research about mathematics teacher preparation and existing standards and to motivate researchers to investigate areas that are less well understood.

AUDIENCE

The audience for these standards includes all those involved in mathematics teacher preparation, including faculty and others involved in the initial preparation of mathematics teachers; classroom teachers and other Pre-K–12 school personnel who support student teachers and field placements; coordinators of mathematics teacher preparation programs; deans, provosts, and other program administrators who make decisions regarding content and funding of mathematics teacher preparation programs; CAEP, the largest accreditor of teacher education programs in the United States as well as state licensure or credentialing agencies/organizations; NCTM, the professional association responsible for setting standards for educator-preparation programs for preservice, middle, and high school mathematics; and other organizations, including specialized professional associations (e.g., NAEYC, CEC) and agencies focused on and involved in the preparation of mathematics teachers.

ORGANIZATION OF THE DOCUMENT

The AMTE Standards for Preparing Teachers of Mathematics are organized in the following way:

- Chapter 1 describes the overall framework, including a set of *assumptions* that underlie the recommendations made throughout the document.
- Chapter 2 provides *standards* for the professional knowledge, skills, and dispositions that well-prepared beginning teachers of mathematics need to possess related to content, teaching, learners and learning, and the social context of mathematics education. Each standard includes *indicators* describing what attainment of that standard by candidates entails.
- Chapter 3 describes *standards* for mathematics teacher preparation programs designed to develop the knowledge, skills, and dispositions of their teacher candidates described in Chapter 2. Again, each standard includes *indicators* of what attainment of that standard by a program entails.
- Chapters 4 through 7 provide specific *elaborations* of the standards in Chapters 2 and 3 to relate them to the specific needs for preparation of teachers of mathematics at different levels of instruction and discuss their alignment with other standards. These grade-bands include Prekindergarten to Grade 2 (early childhood), Grades 3 through 5 (upper elementary), Grades 6 through 8 (middle level), and Grades 9 through 12 (high school).
- Chapter 8 provides *recommendations* for effectively assessing candidates and programs in meeting the standards and elaborations.
- Chapter 9 provides advice on how the vision of this document can be attained, including *action steps* that those involved in mathematics teacher preparation.

CHAPTER 1. INTRODUCTION

As a professional community, mathematics teacher educators have begun to define, research, and refine the characteristics of effective teachers of mathematics and, in particular, the professional proficiencies of a well-prepared beginning teacher of mathematics. This document describes a set of proficiencies for well-prepared beginners and for programs preparing mathematics teachers. Although these proficiencies are grounded in available research, in many areas that research is not yet sufficient to determine the specific knowledge, skills, and dispositions that will enable beginning teachers to be highly effective in their first years of teaching. Hence, the standards presented in this document are intended to engage the mathematics teacher education community in continued research and discussion about what candidates must learn during their initial preparation as teachers of mathematics.

ASSUMPTIONS ABOUT MATHEMATICS TEACHER PREPARATION

The *Standards for Preparing Teachers of Mathematics* are centered on five foundational assumptions about mathematics teacher preparation. These assumptions reflect the emerging consensus of those involved in mathematics teacher preparation in response to the needs of both their teacher candidates and the students those candidates will teach. These assumptions underlie the standards presented in Chapters 2 and 3 as well as the grade-band elaborations in Chapters 4 through 7.

Assumption #1. Ensuring the success of each and every learner requires a deep, integrated focus on equity in every program that prepares teachers of mathematics.

Over the past decades, the need for a central focus on issues related to equity in mathematics education has become clear in reflecting on the uneven performance of students by various demographic factors (AMTE, 2015; NCTM, 2000, 2014a, 2014b). Although equity, diversity, and social-justice issues need to be specifically addressed as standards, they must also be embedded within *all* the standards. Addressing these issues solely within the context of "equity standards" might be misinterpreted to imply that these issues are not important within the other standards; conversely, if they are not directly addressed in standards addressing equity, their centrality to the mission of mathematics teacher preparation can be overlooked. Thus, we believe that equity must be both addressed in its own right and embedded within every standard. Every standard must be built on the premise that it applies to each and every student, recognizing that equity requires acknowledging the particular context, needs, and capabilities of each and every learner rather than providing identical opportunities to students.

Assumption #2. Teaching mathematics effectively requires career-long learning.

Experienced teachers reflecting on their first year of teaching mathematics have frequently described how much more they can now accomplish, given their current level of teaching competence and understanding of the mathematics and students they are teaching. Teachers improve through reflective experience and through intentional efforts to seek additional knowledge. They use that knowledge to build their understanding of the mathematics they teach and to support their improvement in supporting students' learning of mathematics. This process must begin during their initial preparation and continue throughout their careers. Knowing that candidates will complete teacher preparation programs without the expertise they will later develop focuses attention on priorities for beginning teachers. Those priorities become the knowledge, skills, and dispositions of a well-prepared beginner.

Assumption #3. Learning to teach mathematics requires a central focus on mathematics.

Teaching is often approached as a general craft that is independent of the content being taught. Effective mathematics teaching, however, requires not just general pedagogical skills but also content-specific knowledge, skills, and dispositions. To support student learning and develop positive dispositions toward mathematics, mathematics teachers at every level of instruction need deep and flexible knowledge of the mathematics they teach, of how students think about and learn mathematics, of instructional approaches that support mathematical learning, and of the societal context in which mathematics is taught and used in everyday life to effectively support student learning of, and positive dispositions toward, mathematics.

Assumption #4. Multiple stakeholders must be responsible for and invested in preparing teachers of mathematics.

Preparing teachers to teach in ways that ensure that each and every student learns important mathematics requires the concerted effort of everyone who holds a stake in students' future successes. Mathematics teacher educators and mathematicians; other teacher educators; program and school administrators; classroom teachers, including special education teachers; families and communities; policymakers and others in the educational system all play critical roles. When these groups send mixed messages about how mathematics is best taught and learned, beginning teachers receive incomplete and fragmented visions of how to enact effective mathematics learning environments for their students. Successful mathematics teacher preparation requires a shared vision of mathematics learning outcomes for students, of effective mathematics learning environments, and of the kinds of experiences that best support a mathematics teacher's continuing growth and development. Moreover, stakeholders must feel both included in the development of that vision and accountable for enacting that vision.

Assumption #5. Those involved in mathematics teacher preparation must be committed to improving their effectiveness in preparing future teachers of mathematics.

Mathematics teacher preparation program structures differ widely, as do the needs and backgrounds of teacher candidates. Additionally, mathematics teacher preparation occurs within a range of contexts; in the United States, hundreds of institutions as well as online and school district programs prepare teachers of mathematics, and each one is unique. Thus, program personnel need to discern how existing research might apply to their context and how they can respond to issues not yet addressed by research. Further, they must assess the relationship between their unique program and its effectiveness, sharing relevant findings with the broader mathematics teacher preparation community (e.g., through publications and presentations at conferences).

A WELL-PREPARED BEGINNING TEACHER OF MATHEMATICS

As stated in Assumption 2, the development of teachers' content and teaching knowledge, skills, and dispositions occurs over a career-long trajectory. For example, InTASC developed learning progressions to describe "a coherent continuum of expectations for teachers from beginning through accomplished practice" (Council of Chief State School Officers, 2013, p. 6). Figure 1.1 depicts the career-long continuum of teacher development.

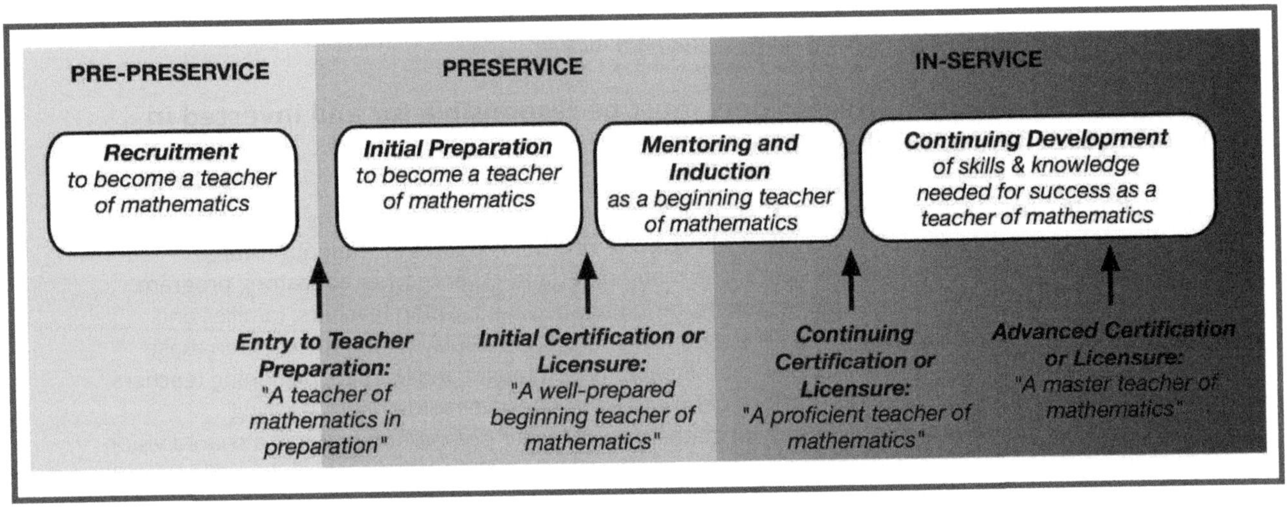

***Figure 1.1.* The teacher development continuum.**

Note. Adapted from *Developing the Analytic Framework: A Tool for Supporting Innovation and Quality Design in the Preparation and Development of Science and Mathematics Teachers* (p. 9) by C. R. Coble, 2012. Washington, DC: Association of Public and Land-grant Universities. Copyright 2012 by APLU.

The standards in this document address primarily the initial preparation phase of the trajectory depicted in Figure 1.1, with some attention to the recruitment of teacher candidates. Chapter 2 provides clear expectations, based on the current knowledge base and national recommendations, for what a well-prepared beginning teacher of mathematics needs to know and be able to do as well as productive dispositions they need to develop, while Chapter 3 describes what programs need to provide to enable candidates to meet these expectations. Well-prepared beginning teachers of mathematics must be committed to supporting the mathematical success of each and every student, and with proper support from the mathematics teacher education community, they will continue to become more effective throughout their careers.

CHAPTER 2. CANDIDATE KNOWLEDGE, SKILLS, AND DISPOSITIONS

Teaching is a complex enterprise, and teaching mathematics is particularly demanding. Thus, that initial preparation focused on teachers' knowledge of the subject, on the curriculum, or on how students learn the subject is more effective than preparation focused on teachers' specific behaviors is not surprising (cf. Ball & Forzani, 2011; Philipp et al., 2007). As described in *Accreditation Standards and Evidence: Aspirations for Educator Preparation* (Council for the Accreditation of Educator Preparation (CAEP), 2013), teacher candidates must learn "critical concepts and principles of their discipline and, by completion, are able to use discipline-specific practices flexibly" (p. 2). This chapter includes *standards* and *indicators* to describe the specific knowledge, skills, and dispositions that well-prepared mathematics teacher candidates at all levels will know and be able to do upon completion of an initial preparation program. Additional expectations specific to a particular grade-band are provided in Chapters 4–7 as *elaborations.* We refer to those who are starting their careers after completion of a teacher preparation program as "well-prepared beginning teachers of mathematics" ("well-prepared beginners" for short).

ORGANIZATION OF THIS CHAPTER

This chapter includes four equally important and interrelated standards that describe the knowledge, skills, and dispositions that well-prepared beginners need to acquire. The first standard, "Knowledge of Mathematics Concepts, Practices, and Curriculum," describes disciplinary knowledge involved in the teaching of mathematics. The second, "Knowledge and Pedagogical Practices for Teaching Mathematics," describes research-based practices or strategies for effective mathematics teaching. The third, "Knowledge of Students as Learners of Mathematics," describes what teachers need to know about their students' mathematical knowledge, skills, representations, and dispositions, for both individual students and groups of students. The final standard in this chapter, "Social Contexts of Mathematics Teaching and Learning," describes the knowledge and dispositions beginning teachers must have related to the social, historical, and institutional contexts of mathematics that affect teaching and learning, themes also woven into the first three standards.

As indicated in Table 2.1, each standard includes specific indicators, along with accompanying explanations. These standards and indicators apply to all well-prepared beginning teachers of mathematics from prekindergarten through high school.

TABLE 2.1. STANDARDS AND RELATED INDICATORS FOR WELL-PREPARED BEGINNING TEACHERS OF MATHEMATICS

STANDARD	RELATED INDICATORS
C.1. Mathematics Concepts, Practices, and Curriculum	
Well-prepared beginning teachers of mathematics possess robust knowledge of mathematical and statistical concepts that underlie what they encounter in teaching. They engage in appropriate mathematical and statistical practices and support their students in doing the same. They can read, analyze, and discuss curriculum, assessment, and standards documents as well as students' mathematical productions.	C.1.1. Know Relevant Mathematical Content C.1.2. Demonstrate Mathematical Practices and Processes C.1.3. Exhibit Productive Mathematical Dispositions C.1.4. Analyze the Mathematical Content of Curriculum C.1.5. Analyze Mathematical Thinking C.1.6. Use Mathematical Tools and Technology
C.2. Pedagogical Knowledge and Practices for Teaching Mathematics	
Well-prepared beginning teachers of mathematics have foundations of pedagogical knowledge, effective and equitable mathematics teaching practices, and positive and productive dispositions toward teaching mathematics to support students' sense making, understanding, and reasoning.	C.2.1. Promote Equitable Teaching C.2.2. Plan for Effective Instruction C.2.3. Implement Effective Instruction C.2.4. Analyze Teaching Practice C.2.5. Enhance Teaching Through Collaboration With Colleagues, Families, and Community Members
C.3. Students as Learners of Mathematics	
Well-prepared beginning teachers of mathematics have foundational understandings of students' mathematical knowledge, skills, and dispositions. They also know how these understandings can contribute to effective teaching and are committed to expanding and deepening their knowledge of students as learners of mathematics.	C.3.1. Anticipate and Attend to Students' Thinking About Mathematics Content C.3.2. Understand and Recognize Students' Engagement in Mathematical Practices C.3.3. Anticipate and Attend to Students' Mathematical Dispositions
C.4. Social Contexts of Mathematics Teaching and Learning	
Well-prepared beginning teachers of mathematics realize that the social, historical, and institutional contexts of mathematics affect teaching and learning and know about and are committed to their critical roles as advocates for each and every student.	C.4.1. Provide Access and Advancement C.4.2. Cultivate Positive Mathematical Identities C.4.3. Draw on Students' Mathematical Strengths C.4.4. Understand Power and Privilege in the History of Mathematics Education C.4.5. Enact Ethical Practice for Advocacy

WHAT SHOULD WELL-PREPARED BEGINNING TEACHERS OF MATHEMATICS KNOW AND BE ABLE TO DO, AND WHAT DISPOSITIONS SHOULD THEY DEVELOP?

The guiding question for this chapter is "Recognizing that learning to teach is an ongoing process over many years, what are reasonable expectations for the most important knowledge, skills, and dispositions that *beginning* teachers of mathematics must possess to be effective?" Answering this question is difficult because some aspects of teaching will not be well learned initially, even though they may be critically important to student learning. What a beginner knows is also a significant equity issue because students with the greatest needs are often taught by teachers with the least experience (Kalogrides, Loeb, & Béteille, 2012; Oakes, 2008).

Mathematics teachers, from the very beginning of their careers, must robustly understand the mathematical content knowledge for the age groups or grades they may teach, along with the content taught to the age groups preceding and following those they teach—and in a different and deeper way than is often presented in textbooks, curriculum documents, or standards. Such knowledge affects their students' learning (e.g., Hill, Rowan, & Ball, 2005; National Mathematics Advisory Panel, 2008).

Well-prepared beginners must be ready to teach each and every student in their first classrooms. Although pedagogical skills develop over time, beginners must have an initial repertoire of effective and equitable teaching strategies; for example, in selecting tasks, orchestrating classroom discussions, building on prior knowledge, and connecting conceptual understanding and procedural fluency (NCTM, 2014a). All teachers, including well-prepared beginners, must hold positive dispositions about mathematics and mathematics learning, such as the notions that mathematics can and must be understood, and that each and every student can develop mathematical proficiency, along with a commitment to imbue their students with similar beliefs and dispositions.

To teach effectively, one must hold knowledge of learners and learning, both general pedagogical knowledge and knowledge specific to the learning and teaching of mathematics. Understanding mathematical learners includes knowing about their backgrounds, interests, strengths, and personalities as well as knowing how students *think about* and *learn* mathematics, including possible misconceptions and creative pathways they may take in learning (Ball & Forzani, 2011; Clements & Sarama, 2014; Sztajn, Confrey, Wilson, & Edgington, 2012). Well-prepared beginners must understand—at least at an initial level—how to assess the understandings and competencies of their students and use this knowledge to plan and modify instruction using research-based instructional strategies (e.g., Ball & Forzani, 2011; Shulman, 1986).

Mathematics teaching and learning are influenced by social, historical, and institutional contexts. Beginning teachers must be aware of learners' social, cultural, and linguistic resources; know learners' histories; and recognize how power relationships affect students' mathematical identities, access, and advancement in mathematics (e.g., Gutiérrez, 2013b; Martin, 2015; Strutchens et al., 2012; Wager, 2012). For example, classroom dynamics and social interactions strongly influence students' emerging mathematical identities, which in turn affect the students' learning opportunities. In short, well-prepared beginners must be ethical advocates for every student.

STANDARD C.1. KNOWLEDGE OF MATHEMATICS FOR TEACHING

Well-prepared beginning teachers of mathematics possess robust knowledge of mathematical and statistical concepts that underlie what they encounter in teaching. They engage in appropriate mathematical and statistical practices and support their students in doing the same. They can read, analyze, and discuss curriculum, assessment, and standards documents as well as students' mathematical productions.

Having a robust knowledge of the mathematics content being taught is foundational to the success of a well-prepared beginning teacher of mathematics. Well-prepared beginners can read, analyze, and discuss curriculum, assessment, and standards documents as well as students' mathematical productions. Without that foundation, they will be unable to support their students' learning of mathematics. As addressed in the preface, in this document the term *mathematics* encompasses mathematics and statistics, because school mathematics teachers are responsible for instruction in both content areas. For cases in which the distinction between mathematics and statistics is important to emphasize, statistics is identified separately.

INDICATOR C.1.1. KNOW RELEVANT MATHEMATICAL CONTENT

Well-prepared beginning teachers of mathematics have solid and flexible knowledge of core mathematical concepts and procedures they will teach, along with knowledge both beyond what they will teach and foundational to those core concepts and procedures.

Well-prepared beginning teachers of mathematics understand and solve problems in more than one way, explain the meanings of key concepts, and explain the mathematical rationales underlying key procedures. For example, a well-prepared beginner for Upper Elementary Grades recognizes that simplifying $3 \div \frac{1}{5}$ indicates the question "How many fifths are in 3?" Using a visual diagram as in Figure 2.1 and considering that one whole is comprised of five fifths leads one to realize that the answer will be 3 × 5 or 15.

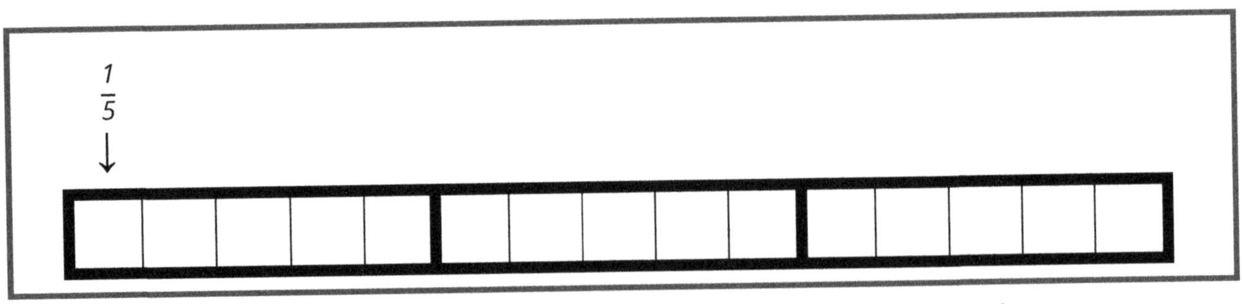

Figure 2.1. **Fraction-bar representation of the problem $3 \div \frac{1}{5}$.**

This result can be generalized so that students recognize that dividing by any unit fraction is equivalent to multiplying by the denominator. Thus, this procedure can be built on a solid and flexible understanding of underlying mathematics. (See Chapters 4 through 7 for additional examples of the specific content for well-prepared beginners at each grade-band.)

Indicator C.1.2. Demonstrate Mathematical Practices and Processes

Well-prepared beginning teachers of mathematics have solid and flexible knowledge of mathematical processes and practices, recognizing that these are tools used to solve problems and communicate ideas.

The mathematical knowledge of well-prepared beginning teachers of mathematics includes ability to use mathematical and statistical processes and practices (NCTM, 2000; NGA & CCSSO, 2010; Shaughnessy, Chance, & Kranendonk, 2009) to solve problems. They use mathematical language with care and precision. These teachers can explain their mathematical thinking using grade-appropriate concepts, procedures, and language, including grade-appropriate definitions and interpretations for key mathematical concepts. They can apply their mathematical knowledge to real-world situations by using mathematical modeling to solve problems appropriate for the grade levels and the students they will teach. They are able to effectively use representations and technological tools appropriate for the mathematics content they will teach. They regard *doing mathematics* as a sense-making activity that promotes perseverance, problem posing, and problem solving. In short, they exemplify the mathematical thinking that will be expected of their students.

Well-prepared beginners recognize processes and practices when they emerge in their mathematical thinking and highlight these actions and behaviors when they observe them in others. Over time, beginning teachers can (a) better distinguish intricacies among the various processes and practices, determining those that are at the crux of a mathematical investigation and (b) see the interrelationships among the processes and practices.

Well-prepared beginners understand that mathematics is a human endeavor that is practiced in and out of school, across many facets of life. They know that mathematics has a history and includes contributions from people with different genders and cultural, linguistic, religious, and racial/ethnic backgrounds. Mathematics is based on constructed conventions and agreements about the meanings of words and symbols, and these conventions vary. Well-prepared beginners are aware that algorithms considered as standard in the United States differ from algorithms used in other countries and that some alternative algorithms have different, desirable properties that make them worth knowing. This idea is elaborated in later *standards* and *indicators* of this chapter as well as in Chapters 4–7.

Indicator C.1.3. Exhibit Productive Mathematical Dispositions

Well-prepared beginning teachers of mathematics expect mathematics to be sensible, useful, and worthwhile for themselves and others, and they believe that all people are capable of thinking mathematically and are able to solve sophisticated mathematical problems with effort.

Well-prepared beginning teachers of mathematics know that one's success in mathematics depends on a productive disposition toward the subject and on hard work (National Research Council [NRC], 2001a). They believe that requisite characteristics of high-quality teaching of mathematics include a commitment to sense making in mathematical thinking, teaching, and learning and to developing *habits of mind,* including curiosity, imagination, inventiveness, risk-taking, and persistence. For example, when faced with a challenge to common practice or to their current understandings or beliefs, well-prepared beginning teachers have the intellectual courage and mathematical disposition to not reject the challenge but to investigate the proposed idea, applying their own critical thinking and using all available resources.

Indicator C.1.4. Analyze the Mathematical Content of Curriculum

Well-prepared beginning teachers of mathematics read, analyze, interpret, and enact mathematics curricula, content trajectories, standards documents, and assessment frameworks for the grades in which they are being prepared to teach.

The alignment of standards, instructional materials, and assessment is critical in designing a cohesive, well-articulated curriculum. Well-prepared beginning teachers of mathematics are aware that the mathematics they teach is based on a variety of, often nested, documents. They know that connections exist among standards, curriculum documents, instructional materials, and assessment frameworks and have dispositions and commitment to analyze these guides to inform their teaching. They have the content preparation and the dispositions to analyze instructional resources, including those provided by textbook publishers and those available from sources online, to determine whether these resources fully address the content expectations described in standards and curriculum documents. When the materials fall short of the standards or expectations, well-prepared beginners are able to decide whether to replace or adapt the materials to better address the content and process expectations.

Well-prepared beginners realize that in addition to the curriculum and standards that they are accountable to teach, other resources can support their efforts to design rigorous, coherent mathematics instruction. Such resources include learning or standards progressions (cf. Generating Increased Science and Mathematics Opportunities, 2012; Institute for Mathematics and Education, n.d.) that describe relationships among standards within and across grades. Note that in addition to content progressions, other types of progressions to be considered are developmental progressions or learning trajectories. Well-prepared beginners understand the content within these materials and can discuss them with colleagues, administrators, and families of their students in ways that make sense to these audiences.

Through analyzing available resources, well-prepared beginners are able to make decisions about the sequencing and time required to teach the content in depth as well as to make important connections among the mathematics taught in the grades or units before and after what they are teaching.

Indicator C.1.5. Analyze Mathematical Thinking

Well-prepared beginning teachers of mathematics analyze different approaches to mathematical work and respond appropriately.

Well-prepared beginning teachers of mathematics analyze both written and oral mathematical productions related to key mathematical ideas and look for and identify sensible mathematical reasoning, even when that reasoning may be atypical or different from their own. Well-prepared beginners value varied approaches to solving a problem, recognizing that engaging in mathematics is more than finding an answer. They make mathematical connections among these approaches to clarify underlying mathematical concepts. Well-prepared beginners recognize the importance of context and applications in uses of mathematics and statistics. They make connections across disciplines in ways that illuminate mathematical ideas.

The tasks in Figures 2.2 and 2.3, which might be used in a mathematics methods or mathematics content course for teachers, exemplify the level of mathematical analysis expected of well-prepared beginners.

Students are given the following prompt:

How many pencils will our classroom need to last through the school year? Propose a strategy to answer this question.

Four student responses follow:

1) We should vote on whether we have enough.

2) We should survey teachers about how many they have used in the past.

3) We should determine how many words each pencil can write and estimate how many words our class will write in a year.

4) We should collect data for a week before trying to answer the question.

What would you ask to further each student's approach to developing a model to answer this question?

Figure 2.2. **Sample task for Pre-K–5 teacher candidates.**

Students are given the following prompt:

One number is 3 times another number, and their sum is 30. What are the two numbers?

Four student responses follow:

1) 30 ÷ 3 = 10, so 10 and 30.

2) 30 ÷ 4 = 7.5, so 7.5 and 22.5.

3) Half of 30 is 15. Half of 15 is 7.5, so go up and down 7.5 from 15.

4) $x + 3x = 30$, so $x = 7.5$.

What question would you ask to clarify each student's thinking?

Figure 2.3. **Sample task for middle level teacher candidates.**

Indicator C.1.6. Use Mathematical Tools and Technology

Well-prepared beginning teachers of mathematics are proficient with tools and technology designed to support mathematical reasoning and sense making, both in doing mathematics themselves and in supporting student learning of mathematics.

Well-prepared beginning teachers of mathematics are proficient in using both digital tools and physical manipulatives for solving mathematical problems and as a means of enhancing or illuminating mathematical and statistical concepts. Well-prepared beginners know when and how to use physical manipulatives to explore mathematical and statistical ideas and to build conceptual understanding of these. Furthermore, they are prepared to use "mathematical action technologies" (cf. NCTM, 2014a, p. 79), powerful tools that will be a part of the lives of the students they teach. They know that physical and digital simulations are critical for understanding key statistical concepts. They are familiar with the use of virtual manipulatives, interactive electronic depictions of physical manipulatives, and know how these can support sophisticated explorations of mathematical concepts (Moyer-Packenham, Niezgoda, & Stanley, 2005).

Well-prepared beginners "make sound decisions about when such tools enhance teaching and learning, recognizing both the insights to be gained and possible limitations of such tools" (NCTM, 2012, p. 3). Not every tool, whether electronic or physical, is appropriate in every situation, and different tools may provide different insights into a context. Well-prepared beginners recognize the fast rate at which technologies emerge and are committed to staying abreast of new tools, analyzing their potential and limitations for students' mathematics learning.

STANDARD C.2. PEDAGOGICAL KNOWLEDGE AND PRACTICES FOR TEACHING MATHEMATICS

Well-prepared beginning teachers of mathematics have foundations of pedagogical knowledge, effective and equitable mathematics teaching practices, and positive and productive dispositions toward teaching mathematics to support students' sense making, understanding, and reasoning.

Teaching mathematics effectively involves significant knowledge, skills, and dispositions. Well-prepared beginning teachers of mathematics draw upon their knowledge of mathematics content, processes, and curriculum and blend it with effective and equitable mathematics teaching practices. They begin teaching with a solid foundation of pedagogical content knowledge. That is, they have studied teaching approaches unique to the subject matter of mathematics, such as knowledge of how to structure and represent mathematics content (e.g., fractions, ratios, algebra) in ways that make it accessible to students; knowledge of the common conceptions, misconceptions, and difficulties that students encounter when learning mathematics; and knowledge of specific teaching strategies to assess and address students' learning needs in mathematics (Shulman, 1986). Well-prepared beginners realize that the teaching of mathematics has its own nuances and complexities specific to the discipline and are ready to start using and expanding their established foundations of pedagogical content knowledge to guide their instructional actions in planning for and interacting with each and every student in mathematics classrooms.

Well-prepared beginners take seriously their responsibilities as new teachers to develop each student's ability to engage in mathematical sense making, understanding, and reasoning. Toward that end, they present positive attitudes toward the subject of mathematics as a discipline when interacting with students, families, colleagues, and community members. They also demonstrate productive dispositions toward teaching mathematics, including engagement in collaborative communities of practice and ongoing learning as professional educators. Well-prepared beginners know that they will continue to develop and refine their pedagogies and practices, and, as with all complex endeavors, their skills and effectiveness in teaching mathematics will improve with time, experience, and deliberate reflection.

Indicator C.2.1. Promote Equitable Teaching

Well prepared beginning teachers of mathematics structure learning opportunities and use teaching practices that provide access, support, and challenge in learning rigorous mathematics to advance the learning of every student.

Teaching for access and equity is evidenced as well-prepared beginning teachers view their roles as developing robust and powerful mathematical identities in their students, demonstrating commitment to view each and every student as a capable and unique learner of mathematics. Well-prepared beginning teachers embrace and build on students' current mathematical ideas and on students' ways of knowing and learning, including attending to each student's culture, race/ethnicity, language, gender, socioeconomic status, cognitive and physical abilities, and personal interests. They also attend to developing students' identities and agency so that students can see mathematics as components of their cultures and see themselves in the mathematics. For instance, well-prepared beginners often connect mathematics to students' everyday experiences or ask students to tell stories because they know that making mathematics relevant to students' lives can make the learning experience more meaningful and raise achievement (Turner, Celedón-Pattichis, Marshall, & Tennison, 2009). Ensuring equitable mathematics learning outcomes for each student is an essential goal and a significant challenge. Achieving this goal requires (a) clear and coherent mathematical goals for students' learning, (b) expectations for the collective work of students in the classroom, (c) effective methods of supporting the learning of mathematics by each student, and (d) provision of appropriate tools and resources targeted to students' specific needs.

Teaching with a commitment to access and equity entails striving to reach each student whose life is affected by what occurs in the mathematics classroom. Well-prepared beginning teachers plan for and use an *equity-based pedagogy* (AMTE, 2015) by structuring learning opportunities to provide access, support, and challenge in learning mathematics. This practice includes considering students' individual needs, cultural experiences, and interests as well as prior mathematical knowledge when selecting tasks and planning for mathematics instruction (Leonard, Brooks, Barnes-Johnson, & Berry, 2010). For example, what everyday, informal language might support or hinder the specialized use of language in mathematics? What are students' prior experiences with specific mathematical representations or strategies? What scaffolds are needed to support students with special needs? What assessments can help identify the strengths and weaknesses of students who are in need of interventions? What topics, activities, places, books, movies, television shows, and video games are currently popular with students and could be incorporated into tasks or problem-solving opportunities?

Well-prepared beginning teachers of mathematics strongly believe that each and every student can learn mathematics with understanding, and they take conscious and intentional actions to build students' agency as mathematical learners (AMTE, 2015; Gutiérrez, 2009). That is, the teachers believe in each student's ability to make sense of mathematical tasks and situations, to engage in mathematical discourse, and to judge the validity of solutions. For example, the beginning teacher envisions a classroom community in which students present ideas, challenge one another, and construct meaning together; varied mathematical strengths are valued and celebrated. Well-prepared beginners understand the importance of communicating the relevance of mathematics—specifically, that students can use mathematics to address problems and issues in their homes and communities.

Additionally, well-prepared beginners intentionally foster growth mindsets among students about learning mathematics and persistently counter manifestations of fixed mindsets (e.g., that some people are good at mathematics and others are not). This practice includes public praise for contributions, use of applicable strategies, and perseverance (Boaler, 2016; Dweck, 2008). For example, well-prepared beginners acknowledge mistakes as critical for learning and help students view mistakes as important in the learning process and for engaging in mathematics.

Indicator C.2.2. Plan for Effective Instruction

Well-prepared beginning teachers of mathematics attend to a multitude of factors to design mathematical learning opportunities for students, including content, students' learning needs, students' strengths, task selection, and the results of formative and summative assessments.

Careful and detailed planning is necessary to design lessons that build on students' mathematical thinking while supporting development of key mathematical ideas. Well-prepared beginners realize that planning takes substantial amounts of time. They recognize the importance of having clear understandings of the mathematics content and mathematics learning goals for each unit and lesson as well as how these particular goals fit within a developmental progression of student learning (Daro, Mosher, & Corcoran, 2011). Well-prepared beginners are able to articulate and clarify mathematics learning goals during the planning process, knowing that the goals are the starting point that "sets the stage for everything else" (Hiebert, Morris, Berk, & Jansen, 2007, p. 57).

To provide effective instruction, one considers the prior knowledge and experiences students bring to a lesson and how the task will be set up or launched in meaningful contexts to ensure that each and every student has access to the content and contexts (Jackson, Garrison, Wilson, Gibbons, & Shahan, 2013). Therefore, well-prepared beginners strive to design classroom environments in which students have opportunities to communicate their thinking, listen to the thinking of others, connect mathematics to a variety of contexts, and make connections across mathematical ideas and subject areas. They plan purposeful and meaningful questions to probe student thinking, make the mathematics visible for discussion, and encourage reflection and justification (NCTM, 2014a).

In effective mathematics planning, teachers select meaningful tasks to motivate student learning, develop new mathematical knowledge, and build connections between conceptual and procedural understanding. Well-prepared beginners understand that providing students opportunities to think, reason, and solve problems requires cognitively challenging mathematical tasks (Stein, Smith, Henningsen, & Silver, 2009). Additionally, well-prepared beginners engage students on a regular basis with mathematical tasks that promote reasoning and problem solving, provide multiple entry points, have high ceilings to offer challenges, and support varied solution strategies (NCTM, 2014a). One can identify such tasks only by considering the ways in which students might solve them. Therefore, well-prepared beginners analyze tasks and lessons, anticipating students' approaches and responses (Gravemeijer, 2004; Stigler & Hiebert, 1999). They understand that anticipating students' responses involves considering both the array of strategies, both conventional and unconventional, that students might use to solve the task, and the ways those strategies relate to the mathematical concepts, representations, and procedures that students are learning (Stein, Engle, Smith, & Hughes, 2008).

Well-prepared beginners attend to the needs of their students in their planning of lessons and units. That is, in their lesson planning, the beginners incorporate inclusive and equity-based teaching practices. Formative assessment is an integral aspect of effective instruction; therefore, lesson planning must include a plan for monitoring and assessing student understanding (Black & Wiliam, 1998). Well-prepared beginners have repertoires of strategies to elicit evidence of students' progress toward the intended mathematics learning goals, such as being able to use observation checklists, interviews, writing prompts, exit tickets, quizzes, and tests. They realize that although they have anticipated student responses, the evidence from these assessments may require a departure from the planned lesson or may affect subsequent lessons within a unit of study. They also recognize their responsibilities to develop interventions to support the students who are not reaching the objectives of the grade-level instruction. Knowing the importance of monitoring and attending to the progress of students who are struggling to learn, the well-prepared beginning teacher uses results from formative assessments to design targeted instruction presented outside the regular grade-level mathematics session.

Indicator C.2.3. Implement Effective Instruction

Well-prepared beginning teachers of mathematics use a core set of pedagogical practices that are effective for developing students' meaningful learning of mathematics.

Teachers must not only understand the mathematics they are expected to teach (Ball, Thames, & Phelps, 2008) and understand how students learn that mathematics (Fuson, Kalchman, & Bransford, 2005), they must be skilled in using content-focused instructional pedagogies to advance the mathematics learning of each and every student (Forzani, 2014). Well-prepared beginning teachers of mathematics have begun to develop skillful use of a core set of effective teaching practices, such as those described in *Principles to Actions* (NCTM, 2014a) and listed in Table 2.2 below.

TABLE 2.2. MATHEMATICS TEACHING PRACTICES (NCTM, 2014a)

Establish mathematics goals to focus learning. Effective teaching of mathematics establishes clear goals for the mathematics that students are learning, situates goals within learning progressions, and uses the goals to guide instructional decisions.

Implement tasks that promote reasoning and problem solving. Effective teaching of mathematics engages students in solving and discussing tasks that promote mathematical reasoning and problem solving and allow multiple entry points and varied solution strategies.

Use and connect mathematical representations. Effective teaching of mathematics engages students in making connections among mathematical representations to deepen understanding of mathematics concepts and procedures and as tools for problem solving.

Facilitate meaningful mathematical discourse. Effective teaching of mathematics facilitates discourse among students to build shared understanding of mathematical ideas by analyzing and comparing student approaches and arguments.

Pose purposeful questions. Effective teaching of mathematics uses purposeful questions to assess and advance students' reasoning and sense making about important mathematical ideas and relationships.

Build procedural fluency from conceptual understanding. Effective teaching of mathematics builds fluency with procedures on a foundation of conceptual understanding so that students, over time, become skillful in using procedures flexibly as they solve contextual and mathematical problems.

Support productive struggle in learning mathematics. Effective teaching of mathematics consistently provides students, individually and collectively, with opportunities and supports to engage in productive struggle as they grapple with mathematical ideas and relationships.

Elicit and use evidence of student thinking. Effective teaching of mathematics uses evidence of student thinking to assess progress toward mathematical understanding and to adjust instruction continually in ways that support and extend learning.

Note. Reprinted from *Principles to Actions: Ensuring Mathematical Success for All* (p. 10) by the National Council of Teachers of Mathematics, 2014. Reston, VA: NCTM. Copyright 2014 by National Council of Teachers of Mathematics. Used with permission.

Well-prepared beginners enter classrooms with commitment to, and initial skills for, enacting effective mathematics instruction. They can identify mathematics learning goals for lessons, articulate how those goals relate to the selected tasks, and use the goals to guide their instructional decisions throughout a lesson. They can distinguish between high-level and low-level tasks on the basis of the cognitive demand for their students. They are growing in their abilities to implement high-level tasks that promote reasoning and problem solving with students and to engage students in meaningful mathematical discourse on those tasks, without lowering the cognitive demand or taking over the thinking and reasoning of students (Stein, Grover, & Henningsen, 1996; Stigler & Hiebert, 2004). For example, well-prepared beginners provide opportunities for students to work together on mathematical tasks and then engage students in whole-class discussions to share, compare, and analyze student strategies and solutions. They endeavor to position students as authors of ideas—students who discuss, explain, and justify their reasoning using varied representations and tools. They also know that the purpose for whole-class discussions is not *show-and-tell* but rather an intentional discussion of selected and sequenced student approaches and use of mathematical representations and tools to move students through a trajectory of sophistication toward the intended mathematics learning goal of the lesson. This skillful orchestration of student interactions in whole-class discussions is complex and takes years for teachers to fully develop. Furthermore, although well-prepared beginners can see key mathematical ideas within students' representations, the accomplished teacher can weave a mathematical idea across many representations (e.g., proportional relationships shown with discrete objects, tables, tape diagrams, double number lines) in ways that help students see connections among representations and see affordances of different representations. Well-prepared beginners have clear visions of effective mathematics instruction centered on mathematical discourse and sense making; they begin with strong foundations of initial capabilities upon which to build that vision into their classrooms.

Effective teaching requires attending to students' mathematical thinking and reasoning during instruction. Well-prepared beginners commit themselves to noticing, eliciting, and using student thinking to assess student progress in understanding the mathematics and to adjust instruction in ways that further support and advance learning toward the intended learning goals. For example, well-prepared beginners assess students' understanding and reasoning at multiple points throughout a lesson. This assessment might include posing purposeful questions to gather information or probe students' understandings while they work individually or in small groups or asking students to respond to a prompt on their whiteboards during a lesson.

Indicator C.2.4. Analyze Teaching Practice

Well-prepared beginning teachers of mathematics are developing as reflective practitioners who elicit and use evidence of student learning and engagement to analyze their teaching.

To effectively reap the benefits of the process of reflection, teachers must base their instructional decisions on evidence of student thinking and reasoning (Wiliam & Leahy, 2015) (See Standard 3.3). Well-prepared beginners analyze the formative assessments used in a lesson to determine both student conceptions and future instruction. They recognize that their analyses must go beyond identifying an overall need and determine precise issues that an intervention could directly support (Hodges, Rose, & Hicks, 2012). For example, a diagnostic interview using a missing-addend problem (e.g., 12 + __ = 23) could reveal a gap in a student's knowledge about the meaning of the equal sign. This information can lead to changes in instruction, such as ensuring that equations are written in a variety of ways in future lessons or the creation of an intervention for Tier 2 instruction incorporating a number balance.

Well-prepared beginning teachers of mathematics recognize that the processes of data collection, analysis, and reflection and the corresponding revision to classroom practices are systematic and continuous and grow in sophistication with teaching experience. Eventually this deliberate examination of practice helps well-prepared beginners become more reflective about their own teaching practices. Various tools enable teachers to design and analyze mathematics teaching. Using these tools, the teacher gathers evidence on students' multiple mathematical knowledge bases and culturally responsive teaching (Aguirre & Zavala, 2013; Roth McDuffie et al., 2014). Well-prepared beginners have the disposition to ask important questions like "How might I get better at this practice I am developing?" and "What other teaching practices might I prioritize?"

Reflecting on one's teaching must not be solely at the individual level. The continuous monitoring of one's practice leads well-prepared beginners to seek out collaborators or critical friends (Schuck & Russell, 2005) to observe one another's teaching, examine students' work samples as a team, and, in concert, consider how particular teaching moves supported or inhibited student understanding and next instructional steps, more specifically. Abandoning carefully designed instructional plans is difficult, but if teaching practices are not working well for students' long-term gains and evidence points to better ways of proceeding, change is necessary. Well-prepared beginners are committed to collaborative analyses of teaching, open to receiving and giving feedback, and capable of making conscious instructional choices on the basis of actual evidence of teaching practice. Changing professional norms encourage teacher-to-teacher collaboration through professional learning communities and formal mentoring and coaching programs. In addition, social media (e.g., blogs; Instagram; Mathematics twitter blogosphere (#MTBoS)) provide virtual spaces for teachers to connect and reflect on their instructional practices.

Indicator C.2.5. Enhance Teaching Through Collaboration With Colleagues, Families, and Community Members

Well-prepared beginning teachers of mathematics seek collaboration with other education professionals, parents, caregivers, and community partners to provide the best mathematics learning opportunities for every student.

Well-prepared beginning teachers of mathematics understand the importance of being a part of a community of educators and believe that the community has the potential to affect teaching in a positive way. "In an excellent mathematics program, educators hold themselves and their colleagues accountable for the mathematical success of every student and for personal and collective professional growth toward effective teaching and learning of mathematics" (NCTM, 2014a, p. 99). Well-prepared beginners anticipate that collaboration with colleagues will spur the need to explain one's teaching practices and articulate rationales for instructional decisions. For example, they are open to making their ideas and decisions visible and subject to shared scrutiny to develop deeper, more widely shared understandings of students' learning (Doerr, Goldsmith, & Lewis, 2010).

Professional learning communities provide teachers with opportunities to collaborate over prolonged periods of time. Five dimensions of successful professional learning communities include (a) common vision and shared values, (b) collective responsibility, (c) leadership support, (d) good facilitation, and (e) use of data and student work (Fulton, Doerr, & Britton, 2010). Observing one another's teaching and providing feedback and protocols for reflecting on practice are often used as key elements in the work of a professional learning community. These dimensions alleviate the isolation of teaching and enable a well-prepared beginner to learn from colleagues and share his or her own expertise. For example, during student teaching and other clinical experiences, well-prepared beginning teachers have had opportunities to observe their peers (i.e., other student teachers) and classroom teachers and reflect on the effectiveness of specific teaching practices. This reflection is scaffolded by university faculty, supervisors, and school-based mentor teachers who provide prospective teachers with writing prompts and oral questions that encourage them to think deeply about what they observed, analyzing both strengths and weaknesses in these situations. Through activities such as these, well-prepared beginning teachers demonstrate a disposition toward teaching as a collaborative endeavor focused on student learning (Leatham & Peterson, 2010).

Engaging with communities outside of school is an important way teachers can strengthen their teaching. Such communities include families, faith-based organizations, public libraries, local businesses, and community centers that provide additional space and mathematical learning opportunities for students, which teachers can leverage (Aguirre et al., 2012; Civil, 2007; González, Andrade, Civil, & Moll, 2001; Vomvoridi-Ivanovic, 2012). Well-prepared beginning teachers are aware of, and committed to employing, multiple strategies to get to know families and communities to better serve students. Examples include annual home/neighborhood visits with families, organizing a robotics or mathematics club at a local library or community center, meeting with a religious or civic leader serving the basic needs of newcomers in the local community, or volunteering in community-based organizations with different language backgrounds to engage beyond their comfort zones.

Well-prepared beginners must be clear and confident in their visions for teaching mathematics. They must be able to effectively communicate their visions while building relationships and trust with families to support mathematics learning throughout the school year. They seek to understand family perspectives and priorities and understand the importance of helping families see how these priorities are being met in their children's mathematics learning. Routine practices such as classroom websites, newsletters, or both; back-to-school nights; family mathematics nights; family surveys; and parent-teacher conferences serve as venues for communicating this vision to multiple stakeholders. That those events include a focus on mathematics, including students' mathematical thinking and how parents and families can support students' mathematics learning, is important.

Well-prepared beginners work diligently to build relationships with families and caregivers focused on the learning needs of their children. They know that providing constructive feedback focused on strengths and areas of growth about students' mathematics performance is essential (Aguirre, Mayfield-Ingram, & Martin, 2013). Important also is to be inclusive when communicating about learning expectations and homework assignments, for example, by translating letters sent home into the parents' languages and honoring mathematics strategies that family members use. Well-prepared beginners are ready with strategies that will ensure that parents understand the rationale for innovations in the teaching and learning of mathematics (e.g., new standards or new teaching approaches) and minimize potential fears and concerns that parents might have about these unfamiliar approaches. These strategies could include organizing family mathematics nights or curriculum nights, sharing specific ways to help with homework, or sending specific family-oriented mathematics activities to parents and caregivers to support their students' learning at home (Hendrickson, Siebert, Smith, Kunzler, & Christensen, 2004; Peressini, 1998). Teachers can also provide resources for family-focused mathematics-project initiatives such as *Family Math* (Stenmark, Thompson, & Cossey, 1986), *Math and Parent Partners (MAPPS)* (Civil & Bernier, 2006), and *Multicultural Literature as a Context for Mathematical Problem Solving: Children and Parents Learning Together* (Strutchens, 2002), and others that have shown how engaging with parents as partners in their students' education can increase their children's achievement and advance equity in mathematics education.

STANDARD C.3. STUDENTS AS LEARNERS OF MATHEMATICS

Well-prepared beginning teachers of mathematics have foundational understandings of students' mathematical knowledge, skills, and dispositions. They also know how these understandings can contribute to effective teaching and are committed to expanding and deepening their knowledge of students as learners of mathematics.

Indicator C.3.1. Anticipate and Attend to Students' Thinking About Mathematics Content

Well-prepared beginning teachers of mathematics anticipate and attend to students' mathematical thinking and mathematical learning progressions.

Understanding how students' mathematical ideas develop and connect is at the core of mathematics teaching. Such understanding rests upon knowledge of the mathematics that comes before and after a given mathematics topic (see Standard C.1.4) as well as understanding of students' informal knowledge, common conceptions, and, for topics for which the research knowledge exists, progressions of students' thinking and learning within the content domain. Well-prepared beginning teachers of mathematics have developed strong understandings of students' mathematical thinking in at least one, and preferably more, well-defined content domain(s) (e.g., within number and operations). They are committed to, and know how to, continue their learning about students' mathematical thinking (e.g., by listening to children and their families, through continued education and professional learning, by using print or online research/resources).

Students come to classrooms with unique mathematical perspectives and experiences. Well-prepared beginners know that the quality and focus of their teaching is affected by the depth and detail of their insight into each student's mathematical thinking. They need to learn about students' informal ideas and invented approaches and students' formal knowledge and understandings as well as how these two types of knowledge influence each other. Understanding the ways that students may think about that mathematical content enables teachers to honor how students' work makes sense to them and to think about instructional moves to use to best extend students' thinking (e.g., students come to question whether their mathematical ideas and procedures make sense in all cases or over different contexts; Jacobs, Lamb, & Philipp, 2010). Well-prepared beginning teachers therefore first try to see mathematical situations through their students' eyes rather than immediately correcting mathematical errors or demonstrating their approaches. This ability to decenter and understand students' thinking is not only useful in planning for instruction but also provides a resource for making sense of moment-to-moment interactions with students.

To understand students' thinking, well-prepared beginners have two competencies. First, they enter classrooms with knowledge of progressions of students' thinking—levels of thinking through which students advance while they learn a specific mathematical topic. They can also access available research-based perspectives on student learning. For example, at early levels of thinking about whole and rational number, students may believe that no numbers exist between two counting numbers or that "multiplication always makes numbers bigger." For beginners, such knowledge is particularly important in content domains prevalent in the curriculum and domains crucial for students' later mathematical success. Well-prepared beginners are disposed to—and have skills that enable them to—learn in an ongoing way about students' ways of thinking related to the mathematical domains at their grade levels, including what comes prior and what will follow.

Second, well-prepared beginning teachers of mathematics can assess and analyze students' thinking. They examine their students' varied approaches to mathematical work and respond appropriately. They gather and use information available through daily classroom interactions, routine formative assessments, summaries documenting students' engagement with computer software or tablet applications, summative assessments, and standardized tests. Well-prepared beginners know the affordances and limitations of these sources of data for understanding student thinking and look for patterns across data sources that provide a sound basis for instructional next steps. They have the mathematical knowledge and the inclination to analyze written and oral student productions, looking for each student's mathematical reasoning even when that reasoning may be different from that of the teacher or the student's peers. They also enhance their own observations by deliberately drawing on the insights of families, professional colleagues, and sources of information from beyond the classroom.

Indicator C.3.2. Understand and Recognize Students' Engagement in Mathematical Practices

Well-prepared beginning teachers of mathematics understand and recognize mathematical practices within what students say and do across many mathematical content domains, with in-depth knowledge of how students use mathematical practices in particular content domains.

Because *doing* mathematics is at the core of learning mathematics, well-prepared beginners recognize that students will present various approaches to problems, representing them, justifying solutions, and critiquing the reasoning of others. As such, they are inclined to anticipate, analyze, and respond to students' diverse solution strategies. Sets of mathematical practices have been usefully elaborated in standards and research syntheses over the past few decades to help teachers know what mathematical practices to develop in their students (cf. NCTM, 2000; National Governors Association Center for Best Practices & Council of Chief State School Officers [NGA & CCSSO], 2010; NRC, 2001a). Although accomplished teachers have greater breadth of knowledge of students' engagement in mathematical practices across many mathematical content domains, well-prepared beginners recognize challenges that students may face when engaging in certain mathematical practices. For example, they know that students may accept arguments that have weak mathematical foundations, especially from peers who are their friends or someone they think typically responds correctly.

Acknowledging that they are only beginning their careers, well-prepared beginners have an emerging understanding of how to nurture development of mathematical processes and practices. They anticipate how students' use of mathematical practices will look and sound within specific grade-band mathematical topics, knowing that over years of experience, their knowledge of students' ways of using mathematical practices will expand to more mathematical topics. For instance, well-prepared middle level teachers can describe how middle level students might reason abstractly and quantitatively about algebra, geometry, and statistics. Well-prepared beginners use their own understanding of mathematical practices responsively when they interact with students who may engage in those practices differently than they do or are just beginning to develop aspects of those practices.

Indicator C.3.3. Anticipate and Attend to Students' Mathematical Dispositions

Well-prepared beginning teachers of mathematics know key facets of students' mathematical dispositions and are sensitized to the ways in which dispositions may affect students' engagement in mathematics.

The ability to engage in mathematics must be coupled with an inclination to see such engagement as worthwhile. Teachers need to know if a student has the disposition "to see sense in mathematics, to perceive it as both useful and worthwhile, to believe that steady effort in learning mathematics pays off, and to see oneself as an effective learner and doer of mathematics" (NRC, 2001a, p. 131). Well-prepared beginners know about these facets of disposition and how to gather information about them from their students. This knowledge reaches beyond determining whether students *like* mathematics and beyond being able to answer students' "When are we going to use this?" questions. Well-prepared beginning teachers recognize that students may attribute mathematical proficiency to innate ability or to the application of effort. They know that these dispositions, although strong at times, are not fixed. They also know the ways in which teachers and schools can inadvertently perpetuate unproductive dispositions. With this knowledge, beginners are prepared to teach in ways that move the student more toward attributing mathematical learning to effort and to extending the meaning and usefulness of mathematics in their students' lives. Finally, well-prepared beginners recognize that disposition is a key area for involving families—in challenging the notion that mathematics is inherited and providing strategies to help their own children develop positive mathematical dispositions.

Well-prepared beginners have ways of learning about, and fostering, the mathematical dispositions of students to include confidence, flexibility, perseverance, curiosity, self-monitoring, and appreciation of mathematics. They recognize the effects of their own language, tone, and expectations in influencing dispositions and mathematical self-images of their students. More than just focusing on the kinds of mindsets that students develop in the mathematics classroom, well-prepared beginners realize that perseverance can be supported when students think that a problem is meaningful. Even though they are well prepared, beginning teachers will not initially know the kinds of variations that exist in students' dispositions. They will not initially have many ways to integrate attention to mathematical dispositions into what often appears to be a school curriculum filled to capacity. With experience, they will develop abilities to refine their approaches to learning about dispositions of the students they teach—such as tailoring their practices to learn about very young students, students with special needs, and students with cultural backgrounds that differ from their own. Well-prepared beginners recognize that such pressures like high-stakes assessment can decrease students' appreciation and confidence and try to minimize such pressures on students.

STANDARD C.4. SOCIAL CONTEXTS OF MATHEMATICS TEACHING AND LEARNING

Well-prepared beginning teachers of mathematics realize that the social, historical, and institutional contexts of mathematics affect teaching and learning and know about and are committed to their critical roles as advocates for each and every student.

Indicator C.4.1. Provide Access and Advancement

Well-prepared beginning teachers of mathematics recognize the difference between access to and advancement in mathematics learning and work to provide access and advancement for every student.

Well-prepared beginners know the meanings of *access* and *advancement,* understanding that denial of access or advancement leads to inequities. Access to mathematics is essential for equitable mathematics education. Access includes ensuring that students have opportunities to learn important mathematics taught by qualified teachers. Well-prepared beginners recognize that access involves structures in schools and in classrooms and recognize classroom practices that threaten access. Access becomes particularly important in the placement of students into higher level courses, in which the focus is on doing mathematics rather than practicing procedures. But, *access* also refers to opportunities within a classroom. Well-prepared beginners realize that access is increased when students can approach a problem from multiple routes (e.g., using a method that is familiar to solve the problem) or when they use curriculum materials that include high-quality, meaningful tasks that go beyond the basic skills often tied to standardized testing. Access is threatened when false hurdles are inherent in the system (e.g., denying access to calculators until students master particular skills such as basic facts).

Advancement is the opportunity to go beyond grade-level expectations to learn additional content. Advancement includes participating in an advanced group in elementary schools, taking algebra prior to ninth grade, taking advanced or college-level courses in high school, and pursuing additional courses or taking honors sections of courses. Well-prepared beginners are prepared to advocate for equitable practices for identifying students for advanced study, recognizing the inadequacy of defining *success* solely by the teacher or standardized tests but also the goals that students hold for themselves, especially students who are Black, Latinx, American Indian, emergent multilinguals, or students of any ethnicity living in poverty (Gutiérrez, Bay-Williams, & Kanold, 2008).

Indicator C.4.2. Cultivate Positive Mathematical Identities

Well-prepared beginning teachers of mathematics recognize that their roles are to cultivate positive mathematical identities with their students.

All mathematics teachers are identity workers in that they contribute to the kinds of identities students develop both inside and outside the classroom (Gutiérrez, 2013b). Students harbor perceptions about what someone who is good at mathematics "looks like" more so than for most subjects; even very young students can identify who in their classrooms are "good" at mathematics, often choosing those who are quick to recall facts or perform algorithms. Well-prepared beginners know that research and standards provide a different description of what being "good at mathematics" entails. For example, *Adding it Up* (NRC, 2001a) described a productive disposition as "the inclination to see mathematics as sensible, useful, and worthwhile, coupled with a belief in diligence and one's own efficacy" (p. 116). Well-prepared beginners seek to actively position all learners as mathematical doers. They understand that developing positive mathematical identities begins with focusing on robust goals for what is important to know and be able to do in mathematics and includes doing mathematics for one's own sake, not just to score well on mathematics tests.

Well-prepared beginners analyze their task selections and implementation, reflecting on the ways they may shape students' mathematical identities. In addition to considering the extent to which the mathematics of the task positions learners as doers, well-prepared beginners consider the contexts of tasks. Contexts such as baseball, rocket launching, or farming may privilege a particular group of students who are familiar and interested in that context. Some classroom practices, such as board races and timed tests, have long-standing history in U.S. classrooms, despite the fact that they exclude those who need more processing time while also communicating that those who are fast are good at mathematics.

Additionally, well-prepared beginners understand that how students' peers and teachers listen to and respect their ideas affects students' emerging mathematical identities (Aguirre et al., 2013). For example, posing higher level questions to students perceived by the teacher as more capable, though perhaps unintentional, can negatively affect the development of positive mathematical identities for those students who are not asked such questions. Similarly, students' failure to attend to one from whom they do not expect a good strategy or to take up the ideas of some group members shapes students' mathematical identities. Issues of power and privilege arise when students judge the validity of mathematical work more on the bases of racial and gender stereotypes of good mathematics students than on the ideas presented (Esmonde & Langer-Osuna, 2013). Well-prepared beginners view their planning, teaching, and assessment as "identity-in-the-making" (Gutiérrez, 2013a, p. 53), resisting explanations that position a student as inferior or on the margins of the classroom culture when they do not participate in the ways expected; these teachers focus instead on how to adapt classroom practices to better support each student's development of a positive mathematical identity.

Indicator C.4.3. Draw on Students' Mathematical Strengths

Well-prepared beginning teachers of mathematics identify and implement practices that draw on students' mathematical, cultural, and linguistic resources/strengths and challenge policies and practices grounded in deficit-based thinking.

When teachers are faced with students who think or speak differently from the mainstream, they may inadvertently seek to remedy those differences rather than seeing them as strengths and resources upon which to build. Educators with deficit-based thinking assume that students are lacking in something; such thinking pervades education policy and practice (Valencia, 2010). Every student enters the classroom with mathematical, cultural, and linguistic strengths that support his or her learning of mathematics. Well-prepared beginners value and notice these *funds of knowledge* (Moll, Amanti, Neff, & Gonzalez, 1992) and draw upon them in ways that help every learner in the classroom. For example, students from other countries are able to offer algorithms that are mathematically correct but unknown to the U.S. students and can underscore the value of different approaches to problem solving or different representations of work (Gutiérrez, 2015; Perkins & Flores, 2002).

Well-prepared beginners also realize that supporting the mathematical learning of emergent multilinguals means attending to language and familiar contexts and experiences that promote conjecturing, reasoning, sense making, and convincing others of mathematical claims so that multilingual students will be encouraged to use their language skills and become valued members of the mathematics classroom (Dominguez, 2011; Moschkovich, 2012). When working with indigenous students, teachers need to consider how a community's needs along with different ways of knowing may influence representations and the reasons for doing mathematics outside of school (Meaney, Trinnick, & Fairhall, 2013; Wagner & Lunney Borden, 2015).

Teachers' beliefs about students can profoundly affect their rationales for student success and failure as well as the decisions teachers make to invest in students' learning. Researchers who ask teachers to assess the abilities of Latinx and Black students generally show significant bias toward negative stereotypes and low expectations (Baron, Tom, & Cooper, 1985; Chval & Pinnow, 2010), and many teachers believe that the achievement gap is at least partially genetic (Bol & Berry, 2005). Well-prepared beginners are prepared to challenge deficit-based thinking in schools and reflect on their own practices in terms of building upon the cultural, linguistic, and unique ways of knowing of their students.

Indicator C.4.4. Understand Power and Privilege in the History of Mathematics Education

Well-prepared beginning teachers of mathematics understand the roles of power, privilege, and oppression in the history of mathematics education and are equipped to question existing educational systems that produce inequitable learning experiences and outcomes for students.

Schools do not exist in a vacuum; administrators and teachers implement policies and practices on the basis of the historical contexts of their schools. Therefore well-prepared beginners must be aware of the national, state, district, and school contexts for educating students and be ready to engage in conversations to address inequitable learning experiences. Well-prepared beginners are cognizant of national reform movements in mathematics education, including the strides and challenges in affording every student a quality mathematics education. For example, well-prepared beginners need to be familiar with current challenges in mathematics education; experts acknowledge that too many students who are Black, Latinx, American Indian, emergent multilingual students, or living in poverty are not being educated well in mathematics. Well-prepared beginners are abreast of professional standards documents such as the *NCTM Standards* (2000, 2014) and *Common Core State Standards–Mathematics* (*CCSS–M*) (NGA & CCSSO, 2010); they realize how these documents have evolved to include more (or less) attention to equitable teaching practices (e.g., principles, position statements). At the state and district levels, a well-prepared beginner seeks to learn about the demographic trends in their districts, segregation and re-segregation policies that result in unequal educational experiences, and the adoption of various textbook and instruction policies. By understanding the historical context of education, well-prepared beginners can learn from past successes and contribute to solving current challenges to advocate for students.

Well-prepared beginners not only know the historical context of mathematics education but also understand that mathematics operates with power and privilege in society. As such, well-prepared beginners are knowledgeable of mathematics education as part of a broader system of mechanisms used to determine what is valued, what is right, and what is normal in society (Valero, 2009). Although expert teachers of mathematics are well grounded in literature that presents strategies for transforming schooling for students who have historically been denied access to a quality mathematics education and implement these practices in their classrooms and schools (e.g., Berry, 2008; Gutstein, 2006; Leonard & Martin, 2013; Rodriguez & Kitchen, 2005), well-prepared beginners have read and know how to access such literature and recognize the importance of implementing practices to empower each and every student. They are prepared to ask questions as needed to understand current policies and practices and to raise awareness of potentially inequitable practices. These practices are particularly important related to students who are Black, Latinx, American Indian, emergent multilingual, or students living in poverty because they are overrepresented in classrooms in which skill-based instruction, worksheets, and computer programs are emphasized instead of problem-solving lessons and high-cognitive-demand tasks (Oakes, 2008). For example, well-prepared beginners might ask how students are recommended and placed in gifted and remedial/intervention programs, whether the placement of students across various programs is representative of the school population, who determines the type of instructional materials that are available to students to support their learning of rigorous mathematics, and how information regarding various mathematics programs is communicated to parents.

Indicator C.4.5. Enact Ethical Practice for Advocacy

Well-prepared beginning teachers of mathematics are knowledgeable about, and accountable for, enacting ethical practices that enable them to advocate for themselves and to challenge the status quo on behalf of their students.

The Council for the Accreditation of Educator Preparation (CAEP, 2013), Mathematics Teacher Education Partnership (2014), and others have cited the importance of teacher dispositions, including ethical practices. To develop into accomplished teachers who affect not just education but also the lives of their students, well-prepared beginners must engage in ethical practices and advocacy. In *ethical practice* professional decisions and actions are guided by a set of principles of mutual respect, integrity, and a sense of justice, not simply external measures of quality teaching like students' test scores or teacher evaluations. Well-prepared beginners exercise ethical practice in decision making and teaching when working with children, families, and colleagues, including the elimination of deficit-based thinking, integration of family/community funds of knowledge into lesson design and implementation, attention to cultural differences, and protection of the rights of privacy.

Well-prepared beginners recognize the need to challenge previous functions of school mathematics and the messages conveyed to students and the greater public through school mathematics. Messages to be challenged include the following: "Being good at mathematics is a sign of intelligence"; "some students are good at mathematics whereas others simply are not"; "mathematics is a natural/pure way of doing things"; "doing mathematics does not involve emotions, values, or the body"; and "all students should aspire to STEM careers." Well-prepared beginners recognize their advocacy roles in teaching, realizing that when teachers fail to take action, students, families, colleagues, and others may be harmed.

Well-prepared beginners demonstrate the understanding that teachers who are successful advocates for students challenge the status quo by developing principled and just ethical practices; they hold themselves and others accountable. Rather than simply reflecting on or analyzing teaching, well-prepared beginners accept that teaching is a political activity (Schoenfeld, 2010) and are prepared to take action in the classroom, at school meetings, in professional settings, with families and the wider community, and in other spaces to advocate for meaningful and robust mathematical student identities and experiences. This knowledge prepares well-prepared beginners to develop language and effective ways of working with allies, choosing their battles appropriately, and being creative and strategic in response to practices and policies that demean students and teachers. Teachers who successfully advocate for students realize that teaching sometimes requires acts of creative insubordination (Gutiérrez, 2015). That is, driven by higher ethics, successful beginning teachers are prepared to re-interpret school rules and practices that are not in the best interests of providing their students meaningful and humane mathematical experiences.

CLOSING REMARKS

In this chapter, we presented four equally important standards (with sets of indicators) to describe the knowledge, skills, and dispositions that well-prepared beginners have upon completion of their preparation programs. These standards are not discrete lists but are interrelated and interdependent. For example, one cannot support the mathematical learning for each student, without having knowledge of content, pedagogical skill, and awareness of social contexts. Setting such high expectations for beginning teacher of mathematics is critical to the success of Pre-K–12 students. For candidates to reach these goals requires strong preparation programs, the focus of Chapter 3.

CHAPTER 3. PROGRAM CHARACTERISTICS TO DEVELOP CANDIDATE KNOWLEDGE, SKILLS, AND DISPOSITIONS

Teacher preparation programs can take various forms, such as undergraduate degree programs in education (or a related field) leading to initial certification or licensure, fifth-year programs (possibly leading to a graduate degree) designed for persons with relevant undergraduate degrees to gain initial certification or licensure, and non-degree programs offered by institutions of higher education, school districts, or other providers designed to enable prospective teachers to attain initial certification or licensure. Moreover, these programs may be specific to mathematics (often focusing on the middle or high school levels) or may lead to more general certifications that include mathematics (such as early childhood or elementary grades, as well as special education). Whatever the structure of the program, it should meet the program standards described in this document to ensure that its completers attain the standards of knowledge, skills, and dispositions needed by a well-prepared beginning teacher of mathematics.

Within this chapter, we often refer to courses. We define *courses* to include university or college courses-for-credit as well as other intensive, ongoing activities that provide equivalent opportunities for professional learning. A learning experience can be considered equivalent to a course, however, only if it includes rigorous goals and objectives and incorporates related assessments to ensure that candidates have met those goals and objectives.

In Chapter 2, we described standards and indicators denoting the knowledge, skills, and dispositions needed by well-prepared beginning teachers of mathematics. In this chapter, we describe standards and indicators for teacher preparation programs to ensure that their students meet those standards. Additional expectations specific to a particular grade-band are provided in Chapters 4 through 7 as elaborations.

ORGANIZATION OF THIS CHAPTER

In this chapter the characteristics of effective mathematics teacher preparation programs are described in five sections. In the first section, the important role of partnerships among stakeholders in ensuring the effective preparation of mathematics teachers is described. In the second, the opportunities to learn mathematics for teaching that need to be provided to candidates are elaborated, whereas in the third, we describe opportunities to learn to effectively teach the mathematics that candidates need. The fourth addresses candidates' opportunities to learn in clinical settings. In the fifth, we discuss effective practices for recruiting (and retaining) candidates into mathematics teacher preparation.

Each standard includes a number of more specific indicators for that standard, along with accompanying explanations. These standards and indicators apply for all well-prepared beginning teachers of mathematics from prekindergarten through high school. Although examples from a particular grade-band may be included to help explain a standard or indicator, Chapters 4 through 7 will provide additional explanation, as appropriate, to address the specialized needs for preparing mathematics teachers at a given grade level.

TABLE 3.1. STANDARDS AND RELATED INDICATORS FOR EFFECTIVE PROGRAMS FOR PREPARING BEGINNING TEACHERS OF MATHEMATICS

STANDARD	RELATED INDICATORS
P.1. Partnerships	
An effective mathematics teacher preparation program has significant input and participation from all appropriate stakeholders.	P.1.1. Engage All Partners Productively P.1.2. Provide Institutional Support
P.2. Opportunities to Learn Mathematics	
An effective mathematics teacher preparation program provides candidates with opportunities to learn mathematics and statistics that are purposefully focused on essential big ideas across content and processes that foster a coherent understanding of mathematics for teaching.	P.2.1. Attend to Mathematics Content Relevant to Teaching P.2.2. Build Mathematical Practices and Processes P.2.3. Provide Sustained, Quality Experiences
P.3. Opportunities to Learn to Teach Mathematics	
An effective mathematics teacher preparation program provides candidates with multiple opportunities to learn to teach through mathematics-specific methods courses (or equivalent professional learning experiences) in which mathematics, practices for teaching mathematics, knowledge of students as learners, and the social contexts of mathematics teaching and learning are integrated.	P.3.1. Address Deep and Meaningful Mathematics Content Knowledge P.3.2. Provide Foundations of Knowledge About Students as Mathematics Learners P.3.3. Address the Social Contexts of Teaching and Learning P.3.4. Incorporate Practice-Based Experiences P.3.5. Provide Effective Mathematics Methods Instructors
P.4. Opportunities to Learn in Clinical Settings	
An effective mathematics teacher preparation program includes clinical experiences that are guided on the basis of a shared vision of high-quality mathematics instruction and have sufficient support structures and personnel to provide coherent, developmentally appropriate opportunities for candidates to teach and to learn from their own teaching and the teaching of others.	P.4.1. Collaboratively Develop and Enact Clinical Experiences P.4.2. Sequence School-Based Experiences P.4.3. Provide Teaching Experiences With Diverse Learners P.4.4. Recruit and Support Qualified Mentor Teachers and Supervisors
P.5. Recruitment and Retention of Teacher Candidates	
An effective mathematics teacher preparation program attracts, nurtures, and graduates high-quality teachers of mathematics who are representative of diverse communities.	P.5.1. Recruit Strong Candidates P.5.2. Address Diverse Community Needs P.5.3. Provide Experiences and Support Structures

STANDARD P.1. PARTNERSHIPS

An effective mathematics teacher preparation program has significant input and participation from all appropriate stakeholders.

The Council for the Accreditation of Educator Preparation (CAEP) (2013) has described the need for and characteristics of effective partnerships. The Association of Public Land-Grant Universities' (APLU) Mathematics Teacher Education Partnership has developed Guiding Principles for Secondary Mathematics Teacher Preparation Programs (Mathematics Teacher Education Partnership, 2014) to provide guidance to educators in partner institutions to educate secondary mathematics teachers. These and other documents can help guide the establishment of the partnership.

Strong partnerships engage all partners in developing a common vision and identifiable goals. As argued by the National Association for Professional Development Schools (2008), a common vision must be focused on "the advancement of the education profession and the improvement of P-12 learning" (p. 3). More specifically, this vision should lead to identifiable goals that promote "professional growth across the continuum of preservice teacher candidates, in-service educators, and college/university faculty and administrators" (p. 4).

While members of the partnership may include a broad range of stakeholders, the partnership *must* include mathematics teacher educators and researchers, their teacher preparation colleagues, Pre-K–12 educators and administrators, mathematicians, statisticians and statistics educators, community-based organizations and community members, and representatives from business and industry. Leadership for creating and sustaining the partnership will typically fall to the mathematics teacher educator.

We envision this group working within a professional council of stakeholders. This council is responsible for reviewing the partnership and verifying that the partnership is inclusive of all stakeholders. Partners recognize that for a collaboration to be successful, it must have systematic ways of verifying that the partnership is mutually beneficial.

Indicator P.1.1. Engage All Partners Productively

An effective teacher preparation program benefits from an interdisciplinary collaborative partnership that is a shared endeavor focused on the preparation of mathematics teachers who are well prepared to improve Pre-K–12 student learning in mathematics from multiple perspectives.

These types of partnerships inform different facets of mathematics teacher preparation based on the wisdom of practice and solid theoretical and research-based knowledge and do so in concert with practitioners to prepare effective mathematics teachers for work in diverse schools.

Mathematics teacher educators (MTEs) provide leadership to ensure that the partners recognize the complex nature of mathematics teaching and learning and respect the contributions needed from each partner. MTEs play a central role in ensuring that the program supports the preparation of well-prepared beginners. The partnership is enhanced by the contributions of research scholars who ensure that the program is designed using the best available knowledge.

The partnership includes active participation by **faculty who teach mathematics and statistics** courses. According to *The Mathematical Education of Teachers II* (*MET II,* CBMS, 2012),"Teacher education should be recognized as an important part of a mathematics department's mission and should be undertaken in collaboration with mathematics education faculty" (p. 19). In addition, "Prospective teachers need mathematics courses that develop a solid understanding of mathematics they will teach;" further, mathematics departments should offer a minimum of "three courses with a primary focus on high school mathematics from an advanced viewpoint" (p. 62). When these criteria are met, an institution's mathematics teacher educators are positioned to partner with mathematics and statistics faculty to ensure that both mathematics courses and mathematics education courses are designed to educate well-prepared beginning teachers. Some students in undergraduate teacher preparation programs begin their higher education at a community college, where they may be taking required mathematics courses for the program. Thus, it is important that partnerships also be formed with mathematics and statistics faculty at community colleges, to ensure that the content courses they offer meet the needs of prospective teachers of mathematics.

An effective partnership requires **collaboration with faculty** in social foundations, special education, educational psychology, educational leadership, and learning technologies as well as faculty in other disciplines who teach courses in teacher preparation, such as statistics educators and engineering educators.

The engagement of **Pre-K–12 school-based personnel** is essential to the partnership. Through shared responsibility, mathematics teacher preparation programs can integrate coursework, theory, and pedagogy. The partnership must ensure that future teachers have high-quality school-based experiences that are needed to educate well-prepared beginners and support the development of candidates' skills as related to the needs of schools and school districts. Similarly, close cooperation with preparation programs helps districts hire teachers who are prepared to be effective in their schools. Building these partnerships not only supports candidates' learning but also deepens classroom teachers' knowledge of mathematics content and pedagogy to support their students' learning.

Families and community leaders are important but often overlooked participants in teacher preparation partnerships. When mathematics teacher educators collaborate with families and leaders in the community, they can design learning experiences that help prepare teachers to better understand family and community cultural perspectives, the various activities and responsibilities students have in their homes and communities, the kinds of mathematics that are performed by community members in their jobs, and the values that are highly regarded. In this way, mathematics teacher educators can help the beginning teacher to reflect on and build lessons and classroom cultures that support students to be themselves and be experts in ways that others in the classroom (including the teacher) may not be.

Members of an effective partnership collaborate with families and community leaders to create activities that are immediately and mutually beneficial for the beginning teacher and students. Such activities could include students' learning mathematics with and from beginning teachers and others in community spaces (e.g., public libraries, Boys & Girls clubs, community centers, or places of worship).

Partnering with **business and industry representatives** helps candidates see the uses of mathematics, technology, and statistics in real-world contexts and ways these partners can assist in supporting students and teachers in learning more about these uses.

Indicator P.1.2. Provide Institutional Support

Institutional commitment for a strong mathematics teacher education program includes institutional support for mathematics teacher educators' career trajectories and appropriate rewards for both their institution and school-based partners.

To have a high-quality teacher preparation program, an institution must provide the resources and the support needed to achieve this vision. In particular, the preparation program's reward structures, including awards, commendations, salaries, and promotion and tenure criteria, must clearly support this work. Mathematics teacher preparation cannot be viewed as a duty or chore delegated to graduate assistants or instructors but rather a core activity of all involved departments.

Institutions must support program-improvement activities, including supporting mathematics teacher educators' attendance at relevant professional conferences so they may learn about other high-quality programs. For example, attending any or all of the following could expand faculty members' awareness of effective program components: the annual meetings of the Association of Mathematics Teacher Educators (AMTE), the Council of Exceptional Children (CEC) Teacher Education Division (TED), and the Professional Development Schools (PDS) conference.

Institutions must provide resources necessary to support teaching and learning. Mathematics teacher educators need access to the materials vital to teaching mathematics in Pre-K–12 schools to prepare beginning teachers for the classrooms in which they will teach. For example, faculty teaching a mathematics methods course should have access to textbooks, online resources, and other instructional materials used by Pre-K–12 teachers in their area to help beginning teachers learn to use curriculum materials effectively.

Education Preparation Providers (EPPs) must value the work involved in developing and improving mathematics teacher education programs. Such work includes communication and collaboration with stakeholders, program-assessment work, and work related to program reviews and meeting program standards. This important work must be valued in the promotion and tenure process.

STANDARD P.2. OPPORTUNITIES TO LEARN MATHEMATICS

An effective mathematics teacher preparation program provides candidates with opportunities to learn mathematics and statistics that are purposefully focused on essential big ideas across content and processes that foster a coherent understanding of mathematics for teaching.

Developing candidates' knowledge of mathematics concepts and practices relevant to teaching must be a priority in an effective teacher preparation program, and developing such knowledge is a career-long endeavor. As addressed in the preface, we use the term *mathematics* to encompass mathematics and statistics because school mathematics teachers are responsible for instruction in both content areas of mathematics and statistics. In instances in which the distinction between mathematics and statistics is important, statistics is identified separately. Within a mathematics teacher preparation program, a variety of individuals provide opportunities for candidates to learn mathematics content. Regardless of their departmental affiliations or academic backgrounds, whether they are employed by a university or a school district, each of these individuals must be considered part of the community of mathematics teacher educators. The responsibility for ensuring appropriate opportunities for candidates to learn mathematics is the joint responsibility of all such mathematics teacher educators.

The NCTM/CAEP Standards (NCTM & CAEP, 2012a, 2012b) describe specific content requirements for secondary mathematics candidates, and *The Mathematical Education of Teachers II (MET II)* (CBMS, 2012) and *Statistical Education of Teachers (SET)* (Franklin et al., 2015) provide specific guidance on the mathematics courses required for teaching at the elementary, middle, and high school levels; these requirements and guidelines are summarized in Table 3.2. **We take these recommendations as the minima for effective mathematics teacher preparation.** More detailed discussion of the mathematics-content expectations is provided in Chapters 4 through 7 of this document.

TABLE 3.2. MINIMUM MATHEMATICS CONTENT PREPARATION FOR TEACHER CANDIDATES

Mathematical Education of Teachers II (2012)	Statistical Education of Teachers (2015)
Pre-K–Grade 5	
Twelve semester-hours of mathematics and statistics courses focused on a careful study of mathematics associated with the *CCSS-M* (K–5 and related aspects of 6–8 domains) (NGA & CCSSO, 2010) from a teacher's perspective.	A minimum of six weeks of instruction included in the coursework.
Grades 6–8	
Mathematics and statistics courses designed for future middle level teachers, including those with a focus on number and operations (6 semester-hours); geometry and measurement (3 semester-hours); algebra and number theory (3 semester-hours); statistics and probability (3 semester-hours); and a further 9 semester-hours of other mathematics and statistics courses selected from available offerings, such as introductory statistics, calculus, number theory, discrete mathematics, history of mathematics, and mathematical modeling.	Two courses in statistics (to be incorporated with courses from *MET II*)
High School	
The equivalent of an undergraduate major in mathematics (including statistics) that includes three courses with a primary focus on high school mathematics from an advanced viewpoint.	Three courses in statistics that include a data-analytic and simulation-based approach with a focus on statistical models and inference (to be incorporated in the undergraduate major)

Indicator P.2.1. Attend to Mathematics Content Relevant to Teaching

An effective mathematics teacher preparation program provides opportunities for candidates to learn, with understanding and depth, the school mathematics and statistics content they will teach.

Opportunities to learn mathematics content needs to be **required of all who are preparing to teach** mathematics or to support student learning of mathematics. That is, in addition being directed to early childhood, upper elementary school, middle and high school mathematics teachers, these standards are directed toward general education teachers at the early childhood and upper elementary level, special education teachers, teachers of emergent multilingual learners, mathematics specialists, and all others who will have responsibility for aspects of student learning in mathematics. However, the preparation needs for these particular specializations vary. We echo the recommendation of the *MET II* report (CBMS, 2012, p. 37): "Special education teachers and ELL teachers who have direct responsibility for teaching mathematics (a core academic subject) should have the same level of mathematical knowledge as general education teachers in the subject." As with all teachers, they must understand not only the mathematics they will teach at a deep level but also mathematics taught before and after that grade (CBMS, 2012). Further details about this recommendation at various grade levels are included in Chapters 4–7.

Programs that are focused on the mathematical content knowledge of beginning teachers of mathematics directly address the issue of equity. All students deserve teachers who possess the content knowledge they need to teach well. By ensuring that those who complete teacher preparation programs have strong content knowledge, understanding of the practice of mathematics, and positive mathematics identities, programs are promoting a teaching workforce that provides an equitable education for each and every student. Of course, this outcome is achievable only if it is accompanied by the perspective that the purpose of teaching mathematics is to meet the needs of each and every learner. Too often, mathematics is used as a sorting subject, one that separates those who can from those who cannot. That belief has a place neither in content courses nor in other content-focused experiences for teachers. Instructors of mathematics and statistics content courses should teach in a way that engenders positive mathematics identities.

Indicator P.2.2. Build Mathematical Practices and Processes

An effective mathematics teacher preparation program provides opportunities for candidates to learn mathematics that enable them to engage in mathematical practices and processes that are appropriate to the content being studied.

Mathematics includes more than learning appropriate content. Effective mathematics teacher preparation programs ensure that candidates are immersed in mathematical practices and processes of reasoning, sense making, and problem solving. Candidates' mathematical experiences include continued emphasis on reasoning abstractly and quantitatively, explaining their thinking, and analyzing the thinking of others. When the content lends itself to such practices and processes, opportunities to learn include seeing structure and generalizing. Mathematical modeling—using mathematics to analyze real-world situations—receives continued attention throughout the program. Programs also provide explicit opportunities for candidates to understand statistical ways of thinking and to understand the range of habits of mind in both mathematical thinking and statistical thinking (Chance, 2002). Programs provide opportunities for candidates to make mathematical connections between various approaches to solving problems and opportunities for candidates to make connections between mathematics and other disciplines. Effective programs provide specific opportunities for candidates to use technology to engage in mathematics and statistics. For example, visualizations of a data set can help students better understand patterns within a data set, and spreadsheets can help students create models of mathematical situations, thus supporting the development of mathematical practices and processes. "Mathematical action technologies" used to "perform mathematical tasks" or "respond to the user's actions in mathematically defined ways" (Dick & Hollebrands, 2011, p. xii) support the development of candidates' deep conceptual understanding through use of mathematical practices and processes.

In effective programs, mathematics content is taught using mathematics teaching methods that serve as models of effective mathematics teaching for candidates. Professional organizations representing the mathematics community recognize the benefits of active learning in mathematics (e.g., CBMS, 2012; MAA, 2015/2016). Active learning in STEM disciplines improves learning (Freeman et al., 2014). In such settings, learners are typically provided challenging tasks that promote mathematical problem solving and are provided opportunities to discuss their thinking in small and full group discourse, thus promoting important mathematical practices (Webb, 2016). Effective programs structure opportunities so that candidates learn mathematics by using such active learning strategies. In this way, candidates experience learning mathematics using methods that are consistent with the methods they should use as teachers.

The instructor's choices about how to address mathematical practices and processes are transparent in effective programs. Often undergraduate students (in this case teacher candidates) focus only on the mathematical products without considering the underlying mathematical practices leading to those products. Effective mathematics instructors emphasize the development of those mathematical practices. In effective programs, mathematics teacher educators explicitly identify and address these mathematical practices for those learning to teach mathematics.

Effective programs focus on building mathematical practices and processes in a manner that honors the social context of teaching and learning mathematics. Those who teach mathematics courses in effective programs model equitable practices, including making apparent underlying beliefs about the role of each individual in the classroom community. For example, in mathematics as a discipline, argument is valued as the way to find mathematical truth, but society does not always value argument from boys and girls equally. Without explicit classroom attention to mathematical argumentation, effective programs risk diminishing the opportunities candidates have to engage in the mathematical practice of constructing viable arguments. Effective programs help teacher candidates who have been successful in the prevailing mathematical culture understand that though mathematics seems to be objective, bias is inherent because it is a human endeavor.

Indicator P.2.3. Provide Sustained, Quality Experiences

An effective mathematics teacher preparation program provides multiple and high-quality opportunities for candidates to learn mathematics relevant to teaching in varied settings across the program.

Mathematics content experiences need to be in-depth, focused, and enacted in varied settings; they should not be diluted, superficial, or limited solely to college classrooms. Assigning one course or even a sequence of courses as an isolated experience does not fulfill the spirit of this standard; engagement with rich content opportunities must be coupled with ongoing integrated learning experiences in mathematics, such as analyzing mathematics content within the context of designing a mathematics lesson for a methods class or clinical experience. Program personnel must acknowledge that deep understanding of mathematics content and practice includes a sustained focus on mathematical ideas throughout all aspects of the program. Mathematics methods or pedagogy courses must be specifically focused on the teaching of mathematics with clear attention to correct mathematics content and productive mathematical practices and processes (i.e., ways of engaging in mathematics). Field experiences and induction experiences must also include a focus on the teaching of important mathematical content, with content-specific rubrics used to evaluate lessons in observations by mentor teachers and university supervisors. As discussed in indicator P.2.2, mathematics content courses should be taught using teaching methods that serve as models of effective instruction. Effective programs instill in beginning teachers the understanding that they will continue to learn mathematics throughout their careers while they make the transition from learning mathematics to joining the profession as teachers of mathematics. Effective programs provide candidates opportunities to learn mathematics by studying mathematics curriculum materials (Davis & Krajcik, 2005). In addition, effective programs provide opportunities for candidates to learn mathematics in collaboration with their colleagues as part of a learning community (NCTM, 2014a).

Opportunities to learn mathematics relevant to teaching may come in mathematics courses and may also be included in other aspects of the program, including clinical experiences, methods courses, involvement in professional mathematics education organizations (local, state, or national), tutoring, service learning, clubs, or research experiences, when these aspects are focused on school mathematics content.

STANDARD P.3. OPPORTUNITIES TO LEARN TO TEACH MATHEMATICS

An effective mathematics teacher preparation program provides candidates with multiple opportunities to learn to teach through mathematics-specific methods courses (or equivalent professional learning experiences) in which mathematics, practices for teaching mathematics, knowledge of students as learners, and the social contexts of mathematics teaching and learning are integrated.

Although many configurations for programs address opportunities to learn to effectively teach mathematics, we use the phrase *mathematics methods courses* to refer to specific types of courses that are neither mathematics courses nor general pedagogy/methods courses but instead lie at the intersection and focus on the pedagogy associated with teaching mathematics. As such, mathematics methods courses must be based on the particular nature of the discipline of mathematics (CBMS, 2012) as well as issues associated with effective teaching of mathematics. In some programs, equivalent opportunities for professional learning may be provided in ongoing activities that include rigorous goals and objectives, and related assessments that ensure that candidates have met those goals and objectives. Opportunities to deeply learn fundamental mathematics simultaneously with issues of pedagogy are critical for prospective mathematics teachers (Steele & Hillen, 2012). The four indicators in this section, although listed separately, should be viewed in an integrated manner as key components of mathematics methods courses.

Indicator P.3.1. Address Deep and Meaningful Mathematics Content Knowledge

An effective mathematics teacher preparation program provides mathematics methods courses or related experiences that represent mathematics as a useful, challenging, and interesting discipline.

Because many candidates enter mathematics methods courses (or equivalent professional learning experiences) without having deeply conceptualized all the mathematics they will be responsible to teach, mathematics methods courses in effective mathematics teacher preparation programs present multiple opportunities for candidates to reconsider and deepen their mathematical understandings as both learners of mathematics and as mathematics teachers. If the approach taken to mathematics is solely or primarily procedural, the other worthy goals of the methods course will be subservient to propagating a widely held practice that mathematics, instead of fundamentally being about engaging in ways of reasoning about quantitative and spatial concepts and principles, is fundamentally about learning to memorize mathematical facts and carry out procedures (Hiebert et al., 2005).

In effective programs, candidates not only learn mathematics concepts and procedures but also develop productive mathematical dispositions. Teacher candidates come with years of experience as mathematical learners and some may hold unproductive beliefs about mathematics and mathematics teaching and learning. Therefore productive mathematical disposition is the oft-missing strand of mathematical proficiency, and mathematics methods courses in effective programs support candidates in developing richer and more positive mathematical dispositions.

Mathematical tasks are central to mathematical reasoning, and rich mathematical tasks emphasizing high cognitive demand (Stein, Smith, Henningsen, & Silver, 2000) must be an integral component of mathematics methods courses. However, even rich mathematical tasks are often rendered devoid of most of their mathematical richness when teachers over-scaffold, thereby leading students through the task without focusing on the underlying thinking and reasoning that the tasks were intended to evoke. In effective programs, candidates in mathematics methods courses not only engage in rich mathematical tasks that are implemented in ways designed to sustain the cognitive demand but also learn to successfully implement high-level tasks in Pre-K–12 settings (Smith, Bill, & Hughes, 2008).

Indicator P.3.2. Provide Foundations of Knowledge About Students as Mathematics Learners

An effective mathematics teacher preparation program provides extensive experiences for candidates to focus on students as mathematics thinkers and learners.

A main learning outcome for a mathematics methods course is to develop candidates' understanding of how students think and learn about mathematics (see Standard C.3). Effective mathematics teacher preparation programs provide multiple experiences (e.g., reading, analyzing videos, conducting teaching experiments) that develop deep understanding of at least one (or two) learning trajectories (e.g., sequences of patterns of thinking for a topic). For example, a methods course would provide extensive, complementary experiences focusing on how students think and learn about a key topic such as whole number and operations for early childhood, fractions for elementary mathematics education, proportional reasoning or algebra for middle school mathematics teachers, and functions for high school mathematics teachers. These experiences also develop competencies in connecting these progressions to specific implications for instruction—that is, candidates learn a complete learning trajectory—and connecting content to mathematical practices.

In effective programs, mathematics methods courses simultaneously provide candidates with experiences to develop strategies for understanding and build students' (a) productive dispositions and positive mathematics identities and (b) meaningful mathematical sense making and use of mathematical practices.

Indicator P.3.3. Address the Social Contexts of Teaching and Learning

An effective mathematics teacher preparation program embeds opportunities for candidates to learn about the social, historical, political, and institutional contexts that affect mathematics teaching and learning. By closely examining these contexts and the structures, policies, and practices that foster and constrain student access to and advancement in mathematics, candidates develop deeper understandings and ethical skill sets for advocacy work in mathematics education.

Mathematics methods experiences provide candidates with foundational knowledge about the social, historical, political, and institutional contexts that affect mathematics teaching and learning. By closely examining these contexts and the structures, policies, and practices that foster and constrain student access to and advancement in mathematics, candidates develop deeper understandings and ethical skill sets for advocacy work in mathematics education.

Many teacher preparation programs include specific courses designed to address social contexts of teaching and learning (e.g. multicultural education), wherein candidates grapple with various equity issues, including examining the roles that power, privilege, and oppression play in schooling (e.g., tracking) as well as effective antiracist and social-justice pedagogies that disrupt institutional bias with teaching innovation, critical reflection, and social action. However, effective mathematics teacher preparation programs also explicitly address equity issues within mathematics methods courses or equivalent professional learning activities that focus specifically on mathematics. For example, mathematics methods experiences might include critical analyses of current mathematics education systems, including the histories and the institutional tools, policies, and practices that shape the mathematics taught, how, and to whom. Such analysis is essential because the current mathematics education system is unjust and grounded in a legacy of segregation, systems of power and privilege, and deficit thinking based on race, ethnicity, class, language, and gender (Berry, Ellis, & Hughes, 2014; Martin, D. B., Gholson, & Leonard, 2010).

Effective programs also help beginning teachers challenge deficit views about learning by questioning the status quo at a systemic level. For example, they consider testing and tracking systems that have instituted ways to identify, label, and separate children by perceived mathematics abilities (Cogan, Schmidt, & Wiley, 2001; Oakes, 2005). Methods courses must provide candidates with tools and frameworks to support a more asset– and resource–based instructional approach focused on students' strengths in learning.

High-quality mathematics teacher preparation programs prepare teachers to navigate the mathematics education political terrain. Teachers often are pressured in the unique high-stakes/high-status context of mathematics education that is consequential to mathematics learning, performance, and student affect. Along with cross-disciplinary coursework, mathematics methods courses provide a foundation for new teachers to recognize, navigate, and begin to understand the challenges associated with ultimately transforming these political contexts into a more just and equitable mathematics education than our nation's youth currently experience (Gutiérrez, 2013a).

Mathematics methods experiences in effective programs prepare beginning teachers to recognize the key roles identity and power play in mathematics education. As identity workers, teachers have tremendous power in how children, their families, and communities see students as doers of mathematics (Gutiérrez, 2013b). Mathematics methods courses offer ways for teacher candidates to critically assess their students' mathematics identities and create learning opportunities to strengthen those mathematics identities in positive ways.

In effective programs, mathematics methods courses provide opportunities for teacher candidates to learn about and build on the multiple mathematical, cultural, linguistic, and family strengths that students bring to the classroom. These activities require the candidates to go beyond traditional field placements and into community settings to learn from and about students, families, and communities. Viewing these social, cultural, and community contexts as resources, rather than barriers, for mathematics teaching and learning requires explicit emphasis in mathematics methods courses (Aguirre et al., 2013; Civil, 2007; Foote, 2009).

Developing an ethical practice for advocacy in mathematics education starts with a strong foundation set in mathematics methods courses and their field experiences. Mathematics methods courses in effective programs provide beginning teachers with opportunities to develop their own stances related to the concept of ethical practice, through such tools as The Mirror Test, a series of critical-reflection questions about ethical obligations to students (Gutiérrez, 2016). Moreover, effective teacher preparation programs assess the ethical practice needed by beginning teachers to inform and improve mathematics instruction.

The development of this ethical practice for advocacy cannot be accomplished in isolation but necessitates collaboration with multiple communities, including face-to-face and virtual communities, to provide candidates needed resources, advice, models, and emotional support to engage in this demanding work. Effective programs provide candidates multiple opportunities to develop knowledge and skills necessary for ethical practices to take action and advocate for students in multiple ways and various settings.

Indicator P.3.4. Incorporate Practice-Based Experiences

An effective mathematics teacher preparation program ensures that practice-based experiences, including mathematics methods courses and equivalent learning experiences, provide candidates with experiences using tools and frameworks grounded in research to develop core pedagogical practices and pedagogical content knowledge for teaching mathematics.

In effective mathematics teacher programs, mathematics methods experiences utilize a practice-based approach (Ball & Cohen, 1999). The decomposition-of-practice movement (Grossman et al., 2009) engages candidates in detailed activities set in the everyday work of teaching. For example, rehearsals (Horn, 2010; Lampert, et al., 2013), launches (Jackson et al., 2013), and student-based experiences ought to play a central role in the mathematics methods class. This practice-based notion of teacher education is different from the traditional view in that teachers experience the emergence of theories from analyses of practice rather than separating the learning theories from the later application of these theories to practice (Smith, 2001).

In mathematics methods experiences, research-based frameworks, tools, and strategies serve as important vehicles for connecting theory and practice and guiding candidates in their work with authentic artifacts of teaching. For example, when focusing attention on key elements of planning a lesson, the methods instructor strategically uses planning tools that emphasize building on students' thinking and orchestrating a productive discussion (Smith et al., 2008), teacher questioning, differentiating instruction, and culturally responsive instruction (Bay-Williams, McGatha, Kobett, & Wray, 2014).

Collaborating with clinical-experience coordinators is an effective way for methods instructors to connect their methods experiences with Pre-K–12 classrooms and children. Professional development schools are an excellent environment for collaboration. Having experiences in clinical settings prior to the student-teaching experience helps teacher candidates situate their learning in practice.

These classroom-based experiences can be enhanced through the careful selection of artifacts of practice, including the use of video and classroom-based cases. Carefully selected videos, paired with focused reflective questions, can change candidates' views of what constitutes effective mathematics instruction (Chval, Lannin, Arbaugh, & Bowzer, 2009). Videos or cases are selected to offer illustrations of practice and learning that capture real teachers teaching real children in real classrooms, including students from various backgrounds, abilities, and understandings engaging in powerful mathematics and utilizing productive strategies, thereby challenging negative racial and gender stereotypes as well as fixed mindsets about who is capable of engaging in rich and rigorous mathematics. When the selection of examples is broadened to classroom-based examples that show students learning mathematics in multiple languages (Celedon-Pattichis & Ramirez, 2012; Moschkovich, 1999, 2002; Vomvoridi-Invanovic & Chval, 2014) and cultural and community experiences (Aguirre & Bunch, 2012), programs make clear statements about not only how mathematics teaching can be effective but also for whom.

Indicator P.3.5. Provide Effective Mathematics Methods Instructors

An effective mathematics teacher preparation program ensures that mathematics methods instructors have relevant grade-level experiences with schools, teachers, and students and possess deep understandings of the mathematics content and the research and practice regarding pedagogical and equity issues associated with effectively teaching each and every student.

Mathematics methods experiences that address the Indicators P.3.1 through P.3.4 provide complex, rich experiences that draw upon a deep view of mathematics and a broad understanding of the mathematics education literature base. Thus, in effective mathematics teacher education preparation programs, instructors for those experiences must possess mathematical knowledge, pedagogical content knowledge, and the knowledge of social-cultural contexts of mathematics. Further, effective methods instructors continually learn from practice and research themselves and encourage candidates to do so throughout their professional journeys (Hiebert et al., 2007). For example, new research about students' thinking in well-defined content domains, advances in educational technology available for mathematics classes, and changing standards require mathematics methods instructors to take the professional stance of viewing teaching as constant learning. Furthermore, given the complex demands of teaching mathematics methods, personnel in effective programs consider how to support mathematics methods instructors while they develop the relevant grade-level school experiences and update their knowledge of research and practice. Also, institutions of higher education that prepare university faculty must explicitly and programmatically support the preparation of methods instructors to meet these expectations.

STANDARD P.4. OPPORTUNITIES TO LEARN IN CLINICAL SETTINGS

An effective mathematics teacher preparation program includes clinical experiences that are guided on the basis of a shared vision of high-quality mathematics instruction and have sufficient support structures and personnel to provide coherent, developmentally appropriate opportunities for candidates to teach and to learn from their own teaching and the teaching of others.

Clinical experiences are guided, hands-on, practical applications and demonstrations of professional knowledge of theory to practice, skills, and dispositions through collaborative and facilitated learning in field-based assignments, tasks, activities, and assessments across a variety of settings. Increased calls to embed clinical experiences in real contexts of teaching illustrate the importance of learning through engagement in teaching (Ball & Forzani, 2009; Grossman et al., 2009; Lampert & Graziani, 2009). Furthermore, clinical experiences along with content knowledge and quality of the candidates, are the components of teacher preparation that are likely to have the strongest effects on outcomes for students (National Research Council [NRC], 2010).

An effective mathematics teacher preparation program provides clinical experiences that are developed mutually with school partners, are scaffolded to build in complexity, include opportunities to work in diverse settings and with a range of learners, and are supervised by qualified mentors. Each of these four critical aspects of clinical experiences is described in this section.

Indicator P.4.1. Collaboratively Develop and Enact Clinical Experiences

An effective mathematics teacher preparation program includes collaboration with school partners to enact a shared vision of effective mathematics teaching.

For candidates to be well prepared in the teaching of mathematics, their programs must provide consistency in what is being taught and modeled in methods courses and field experiences. Effective mathematics teacher preparation programs' personnel work to establish a shared vision and support systems among faculty, supervisors, mentor teachers, and teacher candidates focused on enacting effective mathematics teaching practices (as described in Chapter 2). Through their collaboration, school and university partners develop shared language to discuss teaching and learning as well as co-create needed routines, tools, and norms necessary for achieving that vision.

Beyond being based on shared expectations, effective programs have reciprocal professional relationships among stakeholders (university faculty and supervisors, mentor teachers, and school-based personnel) that are integral to the design, implementation, and ongoing assessment of the preparation program. These relationships might include co-designing field-based assignments, co-teaching methods experiences, or both. In bi-directional partnerships, all aspects of clinical experiences are negotiated with schools and the program provider. For example, effective programs' personnel work collaboratively to select and develop mentor teachers who (a) model effective mathematics teaching and (b) articulate what they are doing and why, coaching the candidates to demonstrate effective teaching.

In a mutually beneficial partnership, clinical experiences are designed to support more than just the candidate or to provide extra classroom support for a teacher. The experience can become a system of simultaneous growth and renewal for the teacher candidate-mentor teacher-university supervisor team when they collaborate; all participants learn and lead while they work on behalf of students. Only when preparation programs purposefully engage *with* schools, not just *in* schools, will their clinical preparation become truly robust in ways that maximize candidates' skill development and therefore their abilities to support the mathematics learning of students.

Indicator P.4.2. Sequence School-Based Experiences

An effective mathematics teacher preparation program supports candidates' engagement in increasingly comprehensive acts of teaching by providing coherent and developmentally appropriate clinical experiences.

The quality of school-based experiences is at least as important as the quantity of time in schools. Observations are ineffective for novices who have not been provided with a critical lens from which to gain insights into teaching. Scaffolded learning experiences of teacher candidates support movement toward classroom independence. Therefore, personnel in an effective mathematics teacher preparation program develop clear trajectories of school-based learning experiences focused on increasingly complex teaching practices (e.g., managing student mathematical discourse moving from individual to small group and then from small group to large group) and on learning through examination of student thinking and instructional practice (Smith, 2001). Mathematics teacher candidates must have multiple opportunities to practice the many skills that define them as well-prepared beginners (see Chapter 2).

An effective mathematics teacher preparation program includes opportunities for teacher candidates to engage in early clinical experiences to begin to shift their lenses from that of a student to that of a teacher and to gain insights into what grade level(s) they would like to teach. In such early experiences, effective programs provide strategies for helping the candidates to develop their mathematics-teaching identities. For example, candidates may be assigned in their field placements to reflect on how teacher-student interactions engage a student in productive struggle, observe or interview a student to understand which mathematical representations the student understands and uses to solve problems, or assist the classroom teacher during individual or small-group work by documenting students' ways of reasoning about the task. These early experiences help candidates determine whether they want to become teachers (and at what level), begin to focus on students' mathematical thinking, and introduce basic ideas about effective instruction. In alternative pathways to teaching, wherein candidates are immediately hired into their own classrooms, strategies such as viewing videos, analyzing vignettes or cases, and comparing student-work samples serve to help candidates focus on student thinking and effective instruction.

When teacher candidates complete their mathematics methods course(s) or related professional learning experiences, their clinical experiences must provide varied and extensive opportunities to connect what they are learning in their coursework to authentic classrooms (see Standard P.3.4). In the case of alternative preparation programs, the candidates are teaching, so they are connecting what they are learning to actions in their own classrooms is even more critical. For example, during a middle school methods course, candidates may solve a proportional-reasoning task, consider ways students may reason about the task, implement the same task with middle school students, then return to the university classroom to discuss what they learned about proportional thinking, students, and their teaching. Candidates need multiple opportunities to teach or co-teach lessons, including opportunities to analyze student work and reflect on the effectiveness of teaching and classroom management in supporting the mathematics learning of each student in a group or class.

Student teaching or internship experiences (in which, over the duration of the placement, candidates take on the full responsibilities of the classrooms) must continue to include a range of activities and assessments that engage the candidate in planning, teaching, assessing, and reflecting on mathematics teaching, at increasingly sophisticated levels. These activities include candidates' focusing on culturally responsive instruction, in particular, identifying teaching practices that support or inhibit the learning of each of their students, noticing that some practices benefit some students but inhibit others. Such reflection by candidates requires significant, focused feedback from mentors who are themselves experienced and skilled at teaching mathematics effectively. Additionally, an effective mathematics teacher preparation program provides opportunities for candidates to co-teach, for example, with their mentor teachers or specialists (e.g., special education teacher or

ESL specialist), reflecting on the ways in which such collaborations can effectively support and challenge each and every student without lowering expectations. Finally, internships provide opportunities for candidates to learn about, participate in, and reflect upon aspects of schooling beyond classroom teaching, such as after-school informal learning, record-keeping, engaging with parents, and analyzing school data.

Effective teacher preparation programs may vary widely in how they sequence experiences, but they have strategically organized a variety of experiences over time such that their candidates are able to develop the myriad of knowledge, skills, and dispositions to enter the workforce well prepared to support students' mathematical engagement and understanding. For example, one field experience model that fosters reflective and student-centered teaching practices is the *paired-placement model* in which two candidates are paired with a single mentor teacher. The mentor teacher provides purposeful coaching and mentoring; and the two candidates offer each other feedback, mentoring, and support (Goodnough, Osmond, Dibbon, Glassman, & Stevens, 2009; Leatham & Peterson, 2010). Another model found to help teacher candidates gain greater pedagogical content knowledge and knowledge of students is *co-planning and co-teaching* (CPCT). CPCT and the paired-placement model promote the collaboration and communication between teacher candidates and mentor teachers who share a common space in the planning, implementation, and assessment of instruction (Bacharach, Heck, & Dahlberg, 2010). Finally, a third model that has shown promise is the *year-long-residency model*, which enables teacher candidates to enact both their practicum and internship experiences in the same classroom. This model incorporates some of the same strategies as the co-planning and co-teaching models and has similar benefits.

Indicator P.4.3. Provide Teaching Experiences With Diverse Learners

An effective mathematics teacher preparation program provides clinical experiences that prepare candidates to teach mathematics to a range of students, in a variety of contexts, and across the grades and content ranges for which they will be certified.

Teachers of mathematics must be prepared to address the academic, socio-emotional, and cultural needs of the diverse students they serve. Field experiences in diverse settings can change teacher candidates' perspectives on cultural diversity and increase teachers' self-efficacy and retention (Castro, 2010; Conaway, Browning, & Purdum-Cassidy, 2007). An effective program ensures that emerging teachers of mathematics have opportunities to work with diverse students. Effective programs also ensure that emerging teachers have significant experiences at the grades and courses for which the candidate is being certified. For example, a program that certifies teachers for Grades 7–12 requires that candidates have significant opportunities to teach diverse learners at both the middle and high school levels.

Placing teacher candidates in a diverse setting is not enough to counteract stereotypical thinking, especially if the experiences do not include opportunities for critical reflection (Bell, Horn, & Roxas, 2007; Garmon, 2004). Effective mathematics teacher preparation programs, therefore, must provide opportunities for candidates to work with a range of students in settings where the mentors and supervisors are able to model and discuss inclusive and culturally responsive mathematics instruction, making visible pragmatic strategies for maintaining high expectations for every student. Some programs are located in less diverse geographical locations, but it is critically important for mathematics teacher candidates in these programs to have opportunities to engage in experiences with diverse learners through other authentic clinical experiences. For example, content-focused pen-pal exchanges and the use of video or live-stream from classrooms provide opportunities for such candidates to explore issues of implicit bias and develop complex understandings of working in culturally and economically diverse settings.

Schools vary greatly in their characteristics, such as the extent to which the mathematics content is integrated, the amount of leveling of students into courses, the philosophy on how mathematics is learned, and the instructional materials (e.g. textbooks, online-curriculum opportunities) that are used. Despite research that tracking or ability grouping can inhibit learning and cause inequities within the system, NAEP data indicate that an increasing number of schools are using these practices (Loveless, 2013). Effective programs, therefore, ensure that candidates have opportunities to teach student groups or classes that are considered below grade-level, at grade-level, and above grade-level, with a focus on the way in which these school structures affect, and perhaps limit, students' opportunities to learn. In an effective program for teachers of mathematics, the fact that students in any track or course deserve high-quality mathematics teaching, through which they are challenged to solve meaningful problems and develop mathematical practices, is emphasized.

Indicator P.4.4. Recruit and Support Qualified Mentor Teachers and Supervisors

An effective mathematics teacher preparation program ensures that mentor teachers and supervisors are able to effectively use clinical settings to support candidates in teaching mathematics well and provide equitable support to each and every student.

Clinical experiences are crucial in supporting the development of beginning teachers who can skillfully do the work of mathematics teaching. Quality clinical experiences provide candidates with scaffolded opportunities to develop skill with teaching practices, insight into mathematics content and into students as learners of that content, and professional orientations and commitments. The quality of these learning opportunities hinges on the support of mentors and teacher preparation supervisors who ensure that candidates are actively leveraging those clinical experiences to learn key knowledge, skills, and dispositions for teaching mathematics (Boyd, Grossman, Lankford, Loeb, & Wyckoff, 2009; Grossman, 2010). In effective programs, mentors and supervisors provide candidates with regular opportunities for critical reflection on mathematics teaching and learning.

Effective mentor teachers know mathematics, model effective mathematics teaching practices, and demonstrate professional commitment to the learning of each and every student. Similarly, supervisors of mathematics candidates have had substantive and successful experience teaching mathematics. They know the mathematics and pedagogical practices that are essential for a well-prepared beginning teacher of mathematics. Further, they know how to utilize the affordances of clinical contexts to support teacher learning. Mentors are willing to use their own teaching as a site for the beginning teacher to learn. They co-plan and co-teach to provide focused support during instruction and regularly provide constructive feedback. Supervisors have command of teacher preparation practices that support the integrated attention to students, content, and teaching practices necessary for skilled engagement in teaching and learning contexts (Lampert et al., 2013).

Effective mentors and supervisors of mathematics teachers promote understanding of the context within schools, helping beginning teachers of mathematics recognize the roles that administrators, parents, and communities play in supporting the mathematics learning of their students. Mentors and supervisors understand the importance of advocating for equitable mathematics learning and can communicate this stance in concrete ways to teacher candidates. For example, a mentor or supervisor might ask a candidate to reflect on how a particular pedagogical move might connect with or affect students' mathematical identities or to reflect on the effects of school structures (e.g., classroom design and resources, tracking or scheduling) on student learning.

An effective mathematics teacher preparation program provides organizational structures that support effective mathematics teaching. They recognize that strong partnerships between teacher preparation programs and schools enhance the instruction of beginning teachers (Grossman, Ronfeldt, & Cohen, 2011; Zeichner & Gore, 1990) by connecting the goals and substance of what candidates are learning through their preparation program and the everyday mathematics learning in Pre-K–12 settings. Programs ensure that those involved with candidates have ongoing professional development in coaching and mentoring to support their skills at supporting candidate learning. Program mentors, supervisors, and instructors have ongoing opportunities to collaborate and communicate to ensure that teacher candidates receive clear and consistent messages about effective mathematics teaching.

STANDARD P.5. RECRUITMENT AND RETENTION OF TEACHER CANDIDATES

An effective mathematics teacher preparation program attracts, nurtures, and graduates high-quality teachers of mathematics who are representative of diverse communities.

A high-quality teacher of mathematics is one who is well prepared to begin teaching mathematics and who is also capable of achieving and exemplifying the knowledge, skills, and dispositions required for meeting the needs of mathematics learners in schools throughout a sustained teaching career. The well-prepared beginning teacher not only has strong mathematics content knowledge but also is skilled at teaching and disposed to care deeply about all students (Wilson & Cooney, 2002). At all levels, the teacher's profile must include caring about children and young adults as well as having a productive disposition toward mathematics learning and teaching. Additionally the profile must include a passion for mathematics and for teaching as well as a commitment to positively affecting the learning of all students (CAEP, 2013). Explicit effort is required in teacher preparation programs to recruit and retain a potential teacher workforce that reflects the diverse communities they will serve. Teacher education programs that address this standard, its indicators below, and the other standards in this document are more likely than others to produce high-quality teacher candidates who will remain in the profession and affect student learning over a career (Darling-Hammond, 2000).

Racial and ethnic demographics of the population of students in today's mathematics classes often differ significantly from demographics of their teachers, who are predominantly female, white, and monolingual (U.S. Department of Education, 2013). The lower college enrollment rates of students of color and barriers such as required scores on licensure examinations limit efforts to increase the diversity of mathematics teachers (Ahmad & Boser, 2014). Even with low enrollments, preparation programs must actively recruit and support a diverse teacher-candidate pool. For example, high-quality teacher candidates' profiles must include evidence of or potential for excellent academic achievement, but the candidates should also demonstrate promise for broader qualities and competencies of effective teachers. Achievements that indicate such promise include working with and meeting the needs of diverse students learning mathematics, particularly those who may have been traditionally discriminated against, excluded, or marginalized. Such experiences are as important to effective recruitment as high performance on traditional measures of academic achievement such as scores on standardized tests.

Indicator P.5.1. Recruit Strong Candidates

An effective mathematics teacher preparation program uses a strategic process for recruiting candidates who are capable of supporting the mathematical learning of each and every student.

The process of recruiting future teachers of mathematics involves a multiple-step approach that includes informing potential candidates about different options and opportunities in teaching, providing experiences in which they work with learners of mathematics, and then assisting them in applying for admission to and seeking financial support for completing a teacher education program. Effective mathematics teacher preparation programs must carefully and strategically manage and involve staff and faculty in all aspects of the recruitment process (Dickey, 2016).

For example, the Mathematics Teacher Education Partnership Research Action Cluster (Ranta & Dickey, 2015) on recruitment has found that strategies specific to attracting high school and college students to programs leading to mathematics certification or licensure include the following:

- Offering field experiences in school mathematics settings with exemplary teachers;
- Providing scholarships targeted to high-need programs;
- Promoting the need for secondary mathematics teachers in ways that demonstrate how the demand exceeds the supply of elementary school teachers or for secondary-level English/language art or social studies teachers;
- Highlighting the integrated and active-learning curriculum intended for elementary and middle level learners;
- Building a connection to the unique emotional and cognitive needs of adolescent learners;
- Providing career counseling to liberal arts majors about major changes and certification options specific to teaching mathematics.

Mathematics teacher preparation programs assess candidates' qualifications for admission using multiple measures that include both cognitive and dispositional factors. This process should uncover candidates' passion for and commitment to mathematics as a discipline and an essential component of an effective citizenry as well as their stances on embracing opportunities to learn, their commitments to helping students grow, and their commitments to equitable teaching. Consistent with CAEP (2013) Standard 2, cognitive measures that include grades in mathematics content courses or standardized mathematics test scores provide valuable information but are not sufficient in making a decision. Identifying dispositions of applicants must be part of an admission interview or essay that the mathematics teacher preparation program uses to inform both admissions and program-planning decisions. For all candidates, prompts that seek to uncover any implicit biases or deficit views of diverse children and families as well as such dispositions to avoid mathematics as "I was never good at mathematics" or "I prefer not to teach mathematics" should be included to allow counseling or a programmatic decision that will improve teacher candidates' attitudes. Additionally, prompts are needed that address a commitment to the needs of all learners and that elicit a commitment to and enthusiasm for teaching and mathematical habits of mind.

Indicator P.5.2. Address Diverse Community Needs

Effective mathematics teacher preparation program personnel actively seek to address the diverse needs of their communities by recruiting future teachers of mathematics using a variety of strategies that include outreach in schools, community colleges, and within the institution.

The diverse needs of communities and locales vary, but mathematics teacher preparation programs' personnel must aspire to produce new teachers whose demographics mirror those of the community they serve (Ahmad & Boser, 2014; Goldhaber & Hansen, 2010). To actively recruit to meet local diversity needs and to address the critical shortage of middle and high school mathematics teachers, the mathematics teacher preparation program should work within middle and high schools, particularly those with clubs or future-teacher groups tied to education careers to build interest in the profession of teaching mathematics through activities like peer tutoring or dual-enrollment courses. These early attempts to reach potential candidates prior to college help plant the seed of being a mathematics teacher as a member of a rewarding profession. Reaching out to community colleges as well as to current STEM majors at the program's institution also provides a means of diversifying the candidate pool. In addition, actively recruiting paraprofessional educators already working in schools, parents or other volunteers, and camp counselors or coaches working in informal educational settings can help diversify a program's applicant pool.

Indicator P.5.3. Provide Experiences and Support Structures

An effective mathematics teacher preparation program provides experiences and support structures to ensure the short- and long-term success of their teacher candidates.

The mathematics teacher preparation program provides appropriate support structures through candidate counseling and advisement, tutoring support in mathematics content, and assistance with Praxis or other related (including state-required) summative- or performance-assessment preparation (Seymour & Hewitt, 1997). Experience and time learning how to access and effectively use support available from other educational professionals and stakeholders, such as special educators, counselors, social workers, and mentors, are critical to retention and success in the profession. Early and frequent field experiences in a variety of settings and targeted experiences across grade levels ensure greater possibilities for success in teaching in both the short- and long-terms (Clift & Brady, 2006). For early childhood, elementary school, and special education teacher candidates, the mathematics teacher preparation program provides strategic placements at varying grade levels to assist in the growth of the candidates' mathematical knowledge for teaching. For secondary programs, the mathematics teacher preparation program systematically diversifies placements to ensure candidates gain experience in diverse courses, grade levels, and settings (CAEP, 2013). At any certification level, field experiences in multiple classrooms and environments of sufficient depth and breadth prior to student teaching should be required (CAEP, 2013). Even when programs' clinical experiences are shorter in duration, as they might be at the post-baccalaureate or graduate levels, collecting information about candidates' past experiences with children and schools allows for targeting placements that complement and diversify the collection of experiences.

CLOSING REMARKS

In this chapter, we described five important standards that are essential aspects of effective mathematics teacher preparation programs. These interrelated standards focus on partnerships, learning mathematics, learning to teach mathematics, learning in clinical settings, and recruitment and retention. High-quality teacher preparation programs attend to each of these important aspects, along with the standards described in Chapter 2, to produce well-prepared beginning teachers of mathematics.

CHAPTER 4. ELABORATIONS OF THE STANDARDS FOR THE PREPARATION OF EARLY CHILDHOOD TEACHERS OF MATHEMATICS

This chapter consists of elaborations and examples of the standards in Chapter 2, describing the knowledge, skills, dispositions, and actions that well-prepared beginning Early Childhood mathematics teachers need to develop, followed by elaborations and examples of the standards in Chapter 3, describing characteristics needed for Pre-K to Grade 2 preservice programs to ensure the effective preparation of their candidates (see Table 4.1). The elaborations in this chapter are focused on those standards involving specific early childhood considerations; therefore, although all the standards in Chapters 2 and 3 apply to the mathematical preparation of early childhood teacher candidates, not all require elaboration.

Although young children are ready and eager to learn mathematics, many early childhood teachers have not been provided substantive opportunities for learning to engage children in rich experiences in domains other than literacy (Institute of Medicine [IOM] & National Research Council [NRC], 2015; NRC, 2001b, 2007). Early childhood educators teach children from birth to age 8 years, an especially critical developmental period of learning for mathematics. These early years form the cognitive foundations of mathematical thinking. Later school success in elementary school depends on preschool children's knowledge of mathematics (Duncan et al., 2007). Further, later reading achievement as well as early reading skills can be predicted from early performance in mathematics (Lerkkanen, Rasku-Puttonen, Aunola, & Nurmi, 2005). The quantitative, spatial, and logical-reasoning competencies of mathematics establish an early cognitive foundation for thinking and learning across subjects.

Given the importance of mathematics to academic success (National Mathematics Advisory Panel, 2008), all children need to develop robust knowledge of mathematics in their earliest years. The development of this knowledge among children as well as the formation of their beliefs and dispositions toward mathematics is dependent upon and related to the capabilities and dispositions of their teachers (Tsamir & Tirosh, 2009). Given the limitations in the present preparation of the early childhood workforce in the domain of mathematics education (IOM & NRC, 2015), that teacher preparation programs attend to the standards set forth in this document is imperative. Effective teacher preparation programs at the early childhood level include focused, sustained, and substantial attention to developing candidates' knowledge of learning trajectories, including deep understanding of early mathematics, knowledge of children's development, and high-leverage pedagogical skills and practices for teaching mathematics. This attention occurs in the context of a comprehensive program of professional preparation for teachers of young children (e.g., addressing social-emotional development, whole child, and curricula that are integrated throughout the day) but with a specific focus on mathematics, as supported by organizations such as the NAEYC (e.g., the standard of "understanding content knowledge and resources in academic disciplines" including mathematics, NAEYC, 2012, p. 36).

TABLE 4.1. ELABORATIONS OF SELECTED CANDIDATE AND PROGRAM STANDARDS FOR EARLY CHILDHOOD TEACHERS OF MATHEMATICS

Effective teacher preparation programs at the early childhood level not only meet the expectations outlined in Chapters 2 and 3 but also demonstrate particular attention to the elaborations of selected standards listed below.

Part 1. Candidate Knowledge, Skills, and Dispositions

EC.1. Deep Understanding of Early Mathematics	Well-prepared beginning teachers of mathematics at the early childhood level have deep understandings of the mathematical concepts and processes important in early learning as well as knowledge beyond what they will teach. [Elaboration of C.1.1]
EC.2. Positive Attitudes Toward Mathematics and Productive Dispositions Toward Teaching Mathematics	Well-prepared beginning teachers of mathematics at the early childhood level demonstrate positive attitudes toward mathematics as a discipline and productive dispositions toward the teaching and learning of mathematics. [Elaboration of C.1.3]
EC.3. Mathematics Learning Trajectories: Paths for Excellence and Equity	Well-prepared beginning teachers of mathematics at the early childhood level understand learning trajectories for key mathematical topics, including how these learning trajectories connect to foundational knowledge, curriculum, and assessment frameworks. [Elaboration of C.1.4]
EC.4. Tools, Tasks, and Talk as Essential Pedagogies for Meaningful Mathematics	Well-prepared beginning teachers of mathematics at the early childhood level intentionally plan for and use tools, tasks, and talk as pedagogies for young children's engagement in meaningful mathematics. [Elaboration of C.2.2 and C.2.3]
EC.5. Understanding Young Children's Mathematical Thinking Informs Teaching	Well-prepared beginning teachers of mathematics at the early childhood level elicit and analyze young children's mathematical thinking to inform classroom interactions and instructional decisions. [Elaboration of C.2.3]
EC.6. Collaboration With Families Enhances Children's Mathematical Development	Well-prepared beginning teachers of mathematics at the early childhood level collaborate with families in a mutually respectful, reciprocal manner to enhance and connect children's in-school and out-of-school mathematical development. [Elaboration of C.2.5]
EC.7. Seeing Mathematics Through Children's Eyes	Well-prepared beginning teachers of mathematics at the early childhood level are conversant in the developmental progressions that are the core components of learning trajectories and strive to see mathematical situations through children's eyes. [Elaboration of C.3.1]
EC.8. Creating Positive Early Childhood Learning Environments	Well-prepared beginning teachers of mathematics at the early childhood level create mathematical learning environments characterized by exploration, reasoning, and problem solving; they draw upon children's mathematical, cultural, and linguistic strengths thereby developing conceptual understanding and positive mathematical identities. [Elaboration of C.4.2 and C.4.3]

Part 2. Program Characteristics	
EC.9. Mathematics Content Preparation of Early Childhood Teachers	Effective programs preparing teachers of mathematics at the early childhood level require at least one mathematics content course (or equivalent professional-learning experiences) focused on key mathematical ideas and processes that are important in early mathematics, including problem solving, number, operations, spatial thinking, shapes, measurement, and early algebraic thinking. [Elaboration of P.2]
EC.10. Mathematics Methods Experiences for Early Childhood Teachers	Effective programs preparing teachers of mathematics at the early childhood level require at least one mathematics methods course (equivalent of 3 semester units) focused on mathematics teaching, children's mathematical thinking, and development of mathematics learning at the early childhood level. [Elaboration of P.2 and P.3]
EC.11. Clinical Experiences in Mathematics for Early Childhood Teachers	Effective programs preparing teachers of mathematics at the early childhood level provide clinical experiences specific to mathematics focused on children's mathematical thinking and mathematics instruction with diverse learners in preschool and primary settings. [Elaboration of P.1 and P.4]

PART 1. ELABORATIONS OF THE KNOWLEDGE, SKILLS, AND DISPOSITIONS NEEDED BY WELL-PREPARED BEGINNING EARLY CHILDHOOD TEACHERS OF MATHEMATICS

This section provides additional detail, commentary, and examples of the knowledge, skills, and dispositions well-prepared early childhood mathematics teachers possess, organized by the general standards in Chapter 2.

STANDARD C.1. MATHEMATICS CONCEPTS, PRACTICES, AND CURRICULUM

Well-prepared beginning teachers of mathematics possess robust knowledge of mathematical and statistical concepts that underlie what they encounter in teaching. They engage in appropriate mathematical and statistical practices and support their students in doing the same. They can read, analyze, and discuss curriculum, assessment, and standards documents as well as students' mathematical productions.

Indicators include

C.1.1. Know Relevant Mathematical Content
C.1.2. Demonstrate Mathematical Practices and Processes
C.1.3. Exhibit Productive Mathematical Dispositions
C.1.4. Analyze the Mathematical Content of Curriculum
C.1.5. Analyze Mathematical Thinking
C.1.6. Use Mathematical Tools and Technology

Effective teachers have deep understandings of the mathematics they are expected to teach and exhibit positive dispositions toward both mathematics teaching and learning. Such understandings and dispositions are particularly critical for early childhood teachers because they develop the foundation of mathematical understanding, beliefs, and attitudes among young learners that start children on their mathematical journeys. Therefore, we have a critical elaboration of this standard for the preparation of early childhood teachers of mathematics.

EC.1. Deep Understanding of Early Mathematics

> Well-prepared beginning teachers of mathematics at the early childhood level have deep understandings of the mathematical concepts and processes important in early learning as well as knowledge beyond what they will teach. [Elaboration of C.1.1]

Although the fact that all teachers need to understand the mathematics they are to teach seems obvious, many have not had experiences in the U.S. educational system to understand key conceptual understandings of the major topics of early mathematics (Ma, 1999; NRC, 2009). Effective teachers at the early childhood level hold deep conceptual understandings of the mathematics they teach as well as knowledge of how these foundational mathematical ideas connect to subsequent learning on the mathematical horizon.

The depth of the mathematics in the early years is often misunderstood (Lee & Ginsburg, 2009):

> Mathematical ideas that are suitable for preschool and the early grades reveal a surprising intricacy and complexity when they are examined in depth. At the deepest levels, they form the foundations of

> mathematics that have been studied extensively by mathematicians over centuries ... and remain a current research topic in mathematics. (NRC, 2009, p. 21)

As an example, teachers need to understand how counting relates to place value, in that only 10 digits are needed to write any counting number by creating larger and larger units (which are the values of places in a written numeral) by taking the value of each place to be equal to 10 of the place to its right (e.g., the *1* in 125 is equal to 10 tens). In this way, every counting number can be expressed in a unique way as a numeral made of a string of digits. These ideas connect the study of counting and place value and are essential in the development of arithmetic, such as for understanding that 63 + 10 is 73 without having to count by ones. Full appreciation of the multiplicative relationship in place value (taking the value of each place to be 10 *times* the value of the previous place to its right), developed over time, is critical for such further mathematics learning as understanding decimal numbers.

High-quality programs prepare candidates with broad and deep understandings of fundamental mathematics. Most important is the domain of number and the related concepts of quantity and relative quantity, counting, and arithmetical operations. Also critical are the domains of geometry and measurement, through which people mentally structure the spaces and objects around them. Connections and coherence among mathematical ideas are enriched when candidates apply number concepts and processes to these spatial structures. In addition, these domains provide rich contexts to further develop the ability to reason mathematically.

In brief summaries below, we highlight the important mathematical concepts and practices that a well-prepared beginning teacher of mathematics must know to be able to support learners in Pre-K through Grade 2. The summaries include specific, but not exhaustive, examples of what understanding and being ready to teach this content entail. These concepts connect to and reflect *The Mathematical Education of Teachers II* (CBMS, 2012) and the research in mathematics teacher education. They are focused on counting and cardinality, number and operations in base ten, operations and algebraic thinking, geometry, measurement, and data. The mathematical ideas at the early childhood level form the subtle and complex foundation of school mathematics. Prospective early childhood teachers and the programs that prepare them well do not dismiss these foundational ideas as simple but rather treat them with the mathematical respect that a careful and sustained examination affords and consider appropriate placement of these topics in either content or methods courses. These ideas are closely connected to their upper elementary successors (as addressed in Chapter 5), and programs preparing early childhood teachers ensure that they offer candidates opportunities to examine the full spectrum of mathematical connections across the preschool through Grade 5 span.

This examination of foundational mathematics includes more than learning appropriate content. Well-prepared early childhood teachers continually engage in the mathematical practices and processes of reasoning, sense making, and problem solving. They use mathematics to model, understand, and analyze real-world situations, and they understand that using manipulatives is not the same as mathematical modeling, which would entail using number sentences to mathematize a real-world, contextual situation (Consortium for Mathematics and Its Applications & Society for Industrial and Applied Mathematics [COMAP & SIAM], 2016).

Counting and Cardinality. Well-prepared beginners understand the fundamental ideas and nuances of counting and cardinality. *Cardinality* means how many things are in a set. Young children determine cardinality by *perceptual subitizing* (immediately recognizing small numbers of objects, up to about 4) or *conceptual subitizing* (using a number composition/decomposition for larger numerosities), counting, or matching. Teachers of mathematics must understand each of these ways, how they are related, and how they develop. All three can be used to establish *quantity*—here, the number of things in a set (another type of quantity is the amount of matter that can be measured—*continuous* quantity). No matter which of the three methods one uses, the number—the cardinality—will be the same. Further, number is an abstraction, and the same number quantifies many different collections, such as three trees, three cookies, or three people. At the core of this commonality is the notion of one-to-one correspondence. Any set of three can be placed in one-to-one correspondence with any other set of three. That is, each member of a set of three cookies can be matched with one and only one member of the set of three people. Thus, any method correctly applied—subitizing, counting, or matching—will result in the same number in a given set. Because of the infinite number list, counting can be used to quantify any discrete set and so is particularly important. Placing the number list in one-to-one correspondence with a set tells how many are in that set; the last number so placed is the cardinal number of the set. We do not have to memorize an infinitely long number list because of our use of the base-ten system, which is discussed below (for more details, see NRC, 2009).

Number and Operations in Base Ten. Well-prepared beginners understand that "comparison of quantities and less-than and greater-than relationships is an early step toward decomposing and composing numbers in ways that are necessary in common addition, subtraction, multiplication, and division procedures" (Dougherty, Flores, Louis, & Sophian, 2010, p. 1). They recognize the importance of the benchmarks of 5 and 10 as support for seeing numbers as combinations of other numbers, such as 7 as the combination of 5 and 2 more; 7 is also 3 away from 10.

Table 4.2 lists items from the *MET II* report (CBMS, 2012) related to number and numeration in Early Childhood.

TABLE 4.2. CONNECTIONS TO *MET II* (CBMS, 2012) RELATED TO NUMBER AND NUMERATION IN EARLY CHILDHOOD

Counting and Cardinality

MET II describes the following essential ideas related to counting and cardinality:

- "The intricacy of learning to count, including the distinction between counting as a list of numbers in order and counting to determine a number of objects." (p. 25)

Number and Operations in Base Ten

MET II describes the following essential ideas related to number and operations in base ten:

- "How the base-ten place-value system relies on repeated bundling in groups of ten and how to use objects, drawings, layered place-value cards, and numerical expressions to help reveal base-ten structure. Developing progressively sophisticated understandings of base-ten structure as indicated by these expressions:

$$357 = 300 + 50 + 7$$

$$= 3 \times 100 + 5 \times 10 + 7 \times 1$$

$$= 3 \times (10 \times 10) + 5 \times 10 + 7 \times 1$$

$$= 3 \times 10^2 + 5 \times 10^1 + 7 \times 10^0.$$

- How efficient base-ten computation methods for addition and subtraction rely on decomposing numbers represented in base ten according to the base-ten units represented by their digits and applying (often informally) properties of operations, including the commutative and associative properties of addition, to decompose the calculation into parts. How to use mathematical drawings or manipulative materials to reveal, discuss, and explain the rationale behind computation methods." (p. 27)

Operations and Algebraic Thinking. Well-prepared beginners need experiences with the varied arithmetic problem types, such as *joining, separating,* and *comparing* problems with different parts of a problem situation unknown (NGA & CCSSO, 2010). For example, a separate, result-unknown is the typical *take-away* problem, but a separate start-unknown is less familiar to many (e.g., "Nita had some cars. She gave three to Kerri and now she has eight. How many did she have to start with?"). Such problems can be used to develop intuitions about the meanings of operations and their relationships to one another. Well-prepared beginners understand the importance of using varied problem situations, recognizing the unique complexities of shifting the location of the unknown and characterizing young children's thinking relative to their solution strategies (Carpenter, Fennema, Franke, Levi, & Empson, 2014). This knowledge enables them to teach in ways that are responsive to their children's thinking while providing a framework to build upon that thinking.

The concepts of addition and subtraction are deeply understood by well-prepared beginning Pre-K to Grade 2 mathematics teachers; they understand multiple representations of the concepts as well as how to sequence and teach this content to students. They recognize the relationship between the content that precedes addition and subtraction (e.g., counting and cardinality) and the content that follows (e.g., multiplication and division).

Well-prepared beginners of early childhood mathematics understand that the "mathematical foundations for understanding computational procedures for addition and subtraction of whole numbers are the properties of addition and place value" (Caldwell, Karp, & Bay-Williams, 2011, p. 28). The commutative and associative properties support flexibility with computations. For example, when asked to find the sum of 4, 5, and 6, one may recognize that 4 and 6 make 10 then combine 10 and 5 to make 15. The associative property allows for the regrouping of addends without changing the sum. Combining the "notion of decomposing a number with the commutative and associative properties is foundational to most addition and subtraction fact strategies" (Caldwell et al., 2011, p. 29).

Finally, well-prepared beginners meaningfully use symbols such as the equal sign. They understand that the *equal sign* denotes that two expressions have the same value, and they avoid the common misconception of the equal sign as merely an indication that the answer comes next (Falkner, Levi, & Carpenter, 1999). Well-prepared beginners challenge children's conceptions by purposefully recording equations that place the equal sign in varied locations, such as 5 = 2 + 3, 2 + 3 = 5, and 5 = 5.

Table 4.3 lists items from the *MET II* report (CBMS, 2012) related to operations and algebraic thinking in early childhood.

TABLE 4.3. CONNECTIONS TO *MET II* (CBMS, 2012) RELATED TO OPERATIONS AND ALGEBRAIC THINKING IN EARLY CHILDHOOD

MET II describes the following essential ideas related to operations and algebraic thinking:

- "The different types of problems solved by addition, subtraction, multiplication, and division, and meanings of the operations illustrated by these problem types.
- Teaching–learning paths for single-digit addition and associated subtraction, including the use of properties of operations....
- Recognizing the foundations of algebra in elementary mathematics, including understanding the equal sign as meaning 'the same amount as' rather than a 'calculate the answer' symbol." (p. 26)

Geometry. Well-prepared beginning teachers of mathematics in Pre-K to Grade 2 possess strong understandings of geometry, including spatial relationships. *Geometry* is the study of shapes and space, including two-dimensional (2D) and three-dimensional (3D) spaces. Although all objects in the world have shape, early 2D geometry is focused on shapes with basic attributes, such as circles, triangles, rectangles, squares, rhombuses, trapezoids, hexagons, and other polygons. Most objects that people make can be modeled with such shapes, especially when they decompose or compose such shapes into more complicated forms. Teachers need to know far more than the names of such shapes. They need to know the components (parts) of the shapes (sides and angles, for examples), the properties of shapes (properties are not components of shapes but rather are the *relationships between two or more components,* such as equal-length sides, parallel sides, or right angles). Thus, the study of geometry is not only naming shapes as wholes; it is also about finding and analyzing the components and properties of shapes. Discussing the *definition* of shapes is a deep and fundamental mathematical practice, and teachers need to understand geometry deeply enough to guide discussions of different informal definitions. For example, beginning teachers need to understand why the all-too-common pseudodefinition of rectangles as having "two long sides and two short sides" is completely inadequate (i.e., many shapes, such as nonrectangular parallelograms fit that description; further, squares do *not* fit that description but *are* rectangles. *Rectangles* are better defined as four-sided polygons with all right angles).

Common 3D shapes include cubes, prisms, cylinders, pyramids, cones, and spheres. Many common objects are approximate versions of these ideal shapes, such as filing cabinets described as rectangular prisms and certain ice cream cones described as cones. As with the study of 2D shapes, the study of 3D shapes is not only about naming these shapes as wholes and learning their names but also entails finding and analyzing their components (e.g., faces, edges, and curved surfaces) and their properties.

Similar to the way that 10 single units can be composed to make a unit of 10, shapes can be composed to make *a unit of unit* shapes or decomposed into smaller shapes. Such compositions and decompositions are important in the study of geometry, and in many other areas in mathematics such as fractions, but also in other subjects, from the sciences to the arts. One special composition is composing squares into rectangular regions—the beginning of spatial structuring of those regions into rows and columns that underlies both coordinate geometry and area measurement. Similarly, composing and decomposing 3D shapes is an important foundation for understanding volume in later grades.

Thus, learning geometry is a central topic for its own sake and for its contributions to learning other topics and other subjects. Further, geometry learning provides opportunities to develop ability to reason mathematically.

Measurement. Well-prepared beginners possess strong understandings of measurement, including quantifying two-dimensional (2D) and three-dimensional (3D) spaces. *Measurement* is the process of determining the size of an object. However, multiple ways exist for determining size, depending on the attribute one chooses. For example, a room may be described as having certain length, width, and height (each 1D) or a certain floor area (2D), or a volume (3D). These measurable geometric attributes are the most important to mathematics (although many more, such as the temperature of the room, are important in science). To measure a quantity, one must choose a unit appropriate to the attribute being measured, and then the size of an object is the number of those units needed to quantify that attribute (e.g., to fill the space or cover the area). Core measurement concepts include *iteration, conservation, and origin*, and provide a framework for connecting linear measurement with measures of area and volume (Clements & Sarama, 2014). (For more details, see NRC, 2009.)

Data. Well-prepared beginners of mathematics in Pre-K to Grade 2 understand that the foundations of statistical reasoning begin with collecting and organizing data to answer a question about our world and then examining the variability of that situation The question is posed, a plan is made to collect data that will address the question, and the data are classified into different categories. The categorized data are usually displayed graphically to describe or compare the categories. Because the process of describing or comparing categories usually involves number or measurement, number and measurement are central to data, and data analysis provides a context in which number and measurement are used (for more details, see National Research Council, 2009).

Table 4.4 lists items from the *MET II* report (CBMS, 2012) related to geometry, measurement, and data in early childhood.

TABLE 4.4. CONNECTIONS TO *MET II* (CBMS, 2012) RELATED TO GEOMETRY, MEASUREMENT, AND DATA IN EARLY CHILDHOOD

Geometry

MET II describes the following essential ideas related to geometry:

- "Understanding geometric concepts of angle, parallel, and perpendicular, and using them in describing and defining shapes; describing and reasoning about spatial locations (including the coordinate plane).
- Classifying shapes into categories and reasoning to explain relationships among the categories." (p. 30)

Measurement and Data

MET II describes the following essential ideas related to measurement and data:

- "The general principles of measurement, the process of iterations, and the central role of units: that measurement requires a choice of measureable attribute, that measurement is comparison with a unit and how the size of a unit affects measurements, and the iteration, additivity, and invariance used in determining measurements.
- How the number line connects measurement with number through length ....
- Using data displays to ask and answer questions about data, and analyzing data with attention to the shape and spread. Examine the way data is [*sic*] collected and what analyses of the data mean about the situation." (p. 29)

EC.2. Positive Attitudes Toward Mathematics and Productive Dispositions Toward Teaching Mathematics

> Well-prepared beginning teachers of mathematics at the early childhood level demonstrate positive attitudes toward mathematics as a discipline and productive dispositions toward the teaching and learning of mathematics. [Elaboration of C.1.3]

For positive dispositions toward mathematics to be cultivated during the mathematical preparation of early childhood teachers is particularly important because teachers' beliefs and affect toward mathematics (see Vignette 4.1[2]) influence what children come to believe and feel toward mathematics (Tsamir & Tirosh, 2009; White, Perry, Way, & Southwell, 2006), and teachers' beliefs influence the opportunities they provide for children to engage in significant mathematical thinking (Staub & Stern, 2002). Children's early experiences with

---

[2] This chapter includes a number of vignettes meant to bring to life the recommendations put forward. The vignettes serve a number of purposes, including proposing tasks that may be used with candidates for particular purposes, providing example interactions from mathematics or mathematics methods courses to exemplify effective instruction, or describing the experiences of a teacher candidate. Each vignette was chosen to highlight a particular point, but use of an isolated vignette may require surrounding context to preserve the spirit intended by the use of the vignette.

mathematics form the foundation for their future success as mathematics learners. That foundation includes not only the development of mathematical knowledge but also the establishment of productive dispositions toward mathematics. *Productive dispositions* are defined as the "habitual inclination to see mathematics as sensible, useful, and worthwhile, coupled with a belief in diligence and one's own efficacy" (NRC, 2001a, p. 116). When teachers make decisions about the mathematical tasks, tools, and discourse within the learning environment, they influence the mathematics content knowledge that children develop as well as children's identities as mathematics learners (Aguirre et al., 2013).

The mathematical preparation of teachers plays a critical role in developing productive beliefs and attitudes toward mathematics (Philipp, 2007; Swars, Smith, Smith, & Hart, 2009). Prospective elementary school teachers' attitudes toward mathematics are affected at two key points (Jong & Hodges, 2015). First, positive shifts in attitudes toward mathematics occur as a result of mathematics methods experiences, and, second, further positive shifts occur as a result of clinical experiences. Throughout the effective mathematical preparation of early childhood teachers, increased attention is given to the development of positive dispositions toward mathematics as a necessary requirement for the teaching of rigorous mathematics, including to our youngest learners in preschool and primary settings.

VIGNETTE 4.1. EXAMINING MEMORIES OF LEARNING MATHEMATICS

Each semester on the first day of my mathematics methods course, I ask my candidates to reflect on their memories of learning mathematics in elementary school, middle school, and high school. For about 10 minutes, they write about memories of their own mathematical experiences (e.g., people, activities, topics, expectations for learning). I encourage them to include not only specific examples of what they remember doing but also descriptions of how they felt in those mathematical situations.

Needless to say, the memories and stories are not all positive. Next the candidates form small groups and share their memories. We then discuss and chart common themes as a whole group, including both positive and negative memories. When memories are shared, I often ask candidates to comment on feelings about themselves as learners of mathematics and about the expectations for understanding or making sense of the mathematics in each situation shared. Unfortunately, over the years, far too few prospective early childhood teachers have shared positive memories. Mathematics was, for many, just something they had to do, and it was not a favorite subject. For others, those reflections evoke bad memories, generate tears, and raise anxieties. Whether the memories were positive or negative, a common theme surfaces that mathematics was for most something to memorize and not something to understand or enjoy.

To close the session, I have the candidates reflect and write once again. This time, I ask them to write about how they would like to be remembered as a teacher of mathematics. As the instructor, I then collect, read, and respond to their writing. This initial experience serves as a backdrop for our work together throughout the semester while each candidate delves more deeply into what being an effective teacher of mathematics at the early childhood level entails.

See Guillaume and Kirtman (2005) and McCulloch, Marshall, DeCuir-Gunby, and Caldwell (2013) for additional resources.

EC.3. Mathematics Learning Trajectories: Paths for Excellence and Equity

> Well-prepared beginning teachers of mathematics at the early childhood level understand learning trajectories for key mathematical topics, including how these learning trajectories connect to foundational knowledge, curriculum, and assessment frameworks. [Elaboration of C.1.4]

Learning trajectories are effective frameworks to use in developing foundations of early mathematics. They are especially important in early childhood for six reasons (Clements & Sarama, 2014). First, children's cognitive development influences how they think and what they can learn about mathematics, arguably, more in these early years than at any other age. Often the variance among children on a variety of cognitive factors is also particularly wide at this level. Second, contexts for teaching (e.g., whole group, learning centers, small groups, individual interactions, informal settings) are varied in early childhood and difficult to coordinate and use for helping children learn mathematics without a teacher's holding a good understanding of learning trajectories. Third, the younger the children the greater the importance of teachers' using children's thinking and prior knowledge as starting points. Fourth, much is known about learning trajectories in *early* mathematics, arguably more than for any other age range (see Table 4.5 for a listing of current research-based learning trajectories). Fifth, despite these reasons, most teachers of young children do not understand or use these learning trajectories. For example, in one study (Supovitz, Ebby, & Sirinides, 2013), only 3% interpreted children's responses at the conceptual level and none at the highest level of the learning trajectories, and more than 3/4 of the teachers suggested only procedural instruction. Sixth, research and developmental work indicate that learning trajectories are effective guides for informing instructional approaches that support young children's learning of mathematics. They also help early childhood educators respect children's developmental processes and constraints as well as their potentials for thinking about and understanding mathematical ideas.

TABLE 4.5. RESEARCH-BASED DEVELOPMENTAL LEARNING TRAJECTORIES

- Recognition of Number and Subitizing
- Verbal and Object Counting
- Comparing, Ordering, and Estimating Numbers
- Early Addition and Subtraction
- Composing Number and Multidigit Addition and Subtraction
- Early Multiplication and Division and Fractions
- Spatial Thinking
- Shapes
- Composition of 3D Shapes
- Composition and Decomposition of 2D Shapes
- Length Measurement
- Area Measurement
- Volume Measurement
- Angle Size

Note. See Clements and Sarama, 2014, and Sarama and Clements, 2009, for detail on each of these trajectories.

Each learning trajectory has three components: a mathematical *goal*, a *developmental progression*, and *instructional strategies* (Sarama & Clements, 2009). To develop a particular mathematical understanding (the goal), children construct each level of thinking and reasoning sequentially (the developmental progression), if provided with appropriate teaching approaches and tasks (the instructional strategies). Prospective early childhood teachers need to learn that these components are intimately and intrinsically interconnected in high-quality instruction. That is, the key to true understanding and successful use of learning trajectories lies not in understanding just each component but in understanding how the components work together and must be used in concert to engage and support young children's learning and thinking about mathematics. This understanding is a demanding expectation, and, thus, we recommend that candidates study in depth at least three research-based developmental learning trajectories. For example, candidates might study learning trajectories for subitizing (shown in Figure 4.1), early adding and subtracting, and length measurement. This in-depth study would include instructional planning and application of all three integrated components for each learning trajectory examined, thus serving as a model for utilization of other development trajectories. In addition, candidates should become familiar with other developmental learning trajectories for early mathematics, including how to find resources to extend their own knowledge and support eventual implementation in their own classrooms.

Goal

The *goal* connects learning trajectories to the important *big ideas* in mathematics. The relevant big idea is that numbers can be used to tell how many, describe order, and measure; they involve numerous relations and can be represented in various ways. The specific goals include understanding that subitizing can be used to tell how many (perceptual subitizing), that a quantity can consist of parts and can be broken apart (decomposed) into the parts, and that those parts can be combined (composed) to form the whole (conceptual subitizing). Specific objectives are for children to achieve those understandings and fluency in perceptual and conceptual subitizing.

Age (yrs)	Developmental Progression	Sample Instructional Tasks
0–1	**Pre-Explicit Number.** Within the first year, the child is sensitive to small numbers, but does not have explicit, intentional knowledge of number.	*Noticing Collections*. Provide a rich sensory environment, use words such as *more,* and use actions of adding objects.
1–2	**Small-Collection Namer**. Names groups of 1 to 2, sometimes 3. Shown a pair of shoes, says, "Two shoes."	*Board Games–Small Numbers*. Play board games with a special die (number cube) or spinner that shows only 1, 2, and 3 dots (then add 0 to it).
3	**Maker of Small Collections**. Nonverbally makes a small collection (no more than 4, usually 1-3) with the same number as another collection via mental model (i.e., not necessarily by matching). Might also be verbal. When shown a collection of 3, makes another collection of 3.	*Get the Number*. Ask children to get the right number of crackers or some other item for a small number of children.
4	**Perceptual Subitizer to 4.** Instantly recognizes collections up to 4 briefly shown, and verbally names the number of items. When shown objects briefly, says, "Four."	*Snapshots*. Play "Snapshots" with collections of one to four objects, arranged in line or other simple arrangement, asking children to respond verbally with the number name. Start with the smaller quantities and easier arrangements, moving to those of moderate difficulty only when children are fully competent and confident. 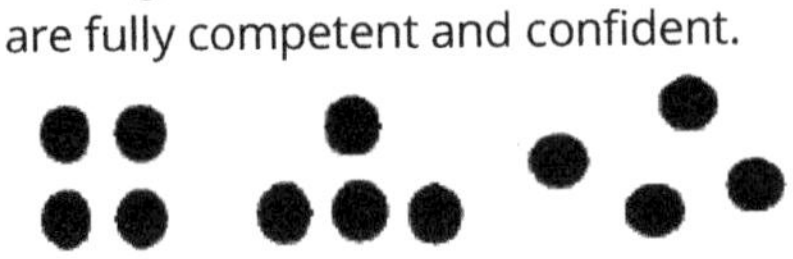 easy medium difficult
5	**Perceptual Subitizer to 5.** Instantly recognizes briefly shown collections up to 5 items and verbally names the number of items. Recognizes and uses spatial and numeric structures beyond the situations in which they were already experienced. Shown 5 objects briefly, says, "Five."	*Snapshots*: Play **"**Snapshots**"** with dot cards, starting with easy arrangements, moving to more difficult arrangements when children are able to do so. 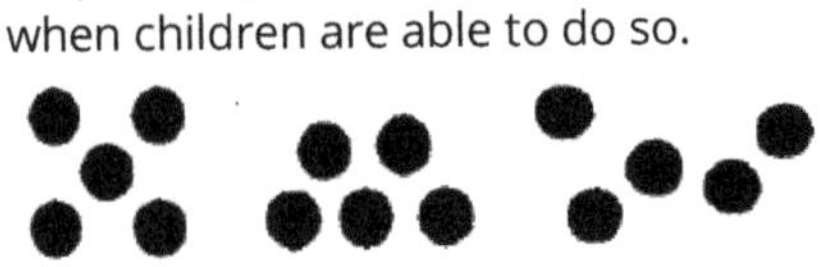 easy medium difficult

5	**Conceptual Subitizer to 5**. Verbally labels all arrangements to about 5, when shown only briefly. Asked "Why?" says, "Because I saw 3 and 2, and so I said 5."	*Snapshots*. Use different arrangements of the various modifications of "Snapshots" to develop conceptual subitizing and ideas about addition and subtraction. The goal is to encourage children to see the addends and the sum. 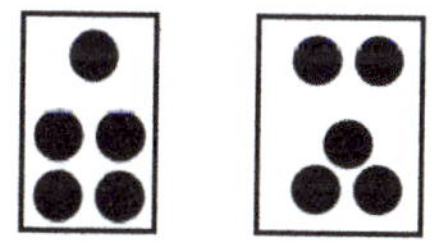
5	**Conceptual Subitizer to 10**. Verbally labels most briefly shown arrangements to 6, then up to 10, using groups. "In my mind, I made two groups of 3 and one more, so 7."	*Snapshots*. Play "Snapshots" with larger quantities to develop ideas about addition and subtraction. 
6	**Conceptual Subitizer to 20**. Verbally labels structured arrangements up to 20, shown only briefly, using groups. "I saw three 5s, so 5, 10, 15."	*Ten Frame Addition Snapshots*: Briefly show 2 ten frames to help children visualize addition combinations. 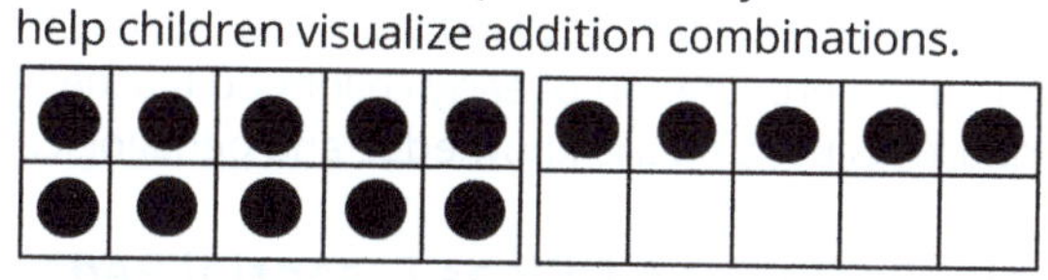
7	**Conceptual Subitizer With Place Value and Skip Counting**. Verbally labels structured arrangements, shown only briefly, using groups, skip counting, and place value. "I saw groups of 10s and 2s, so 10, 20, 30, 40, 42, 44,. . . 44!"	*Ten Frame Addition Snapshots*. Briefly show several ten frames. 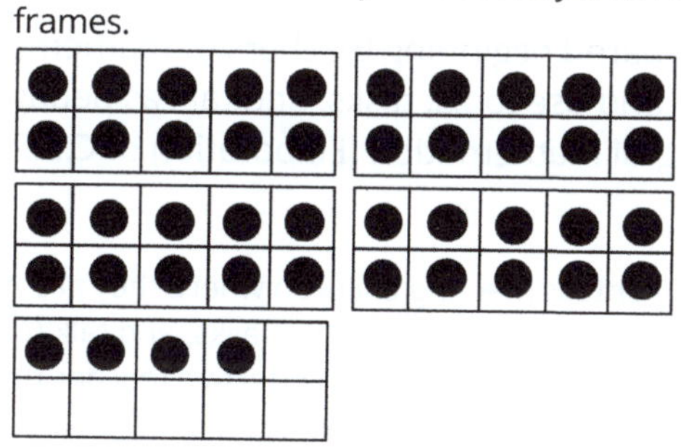
8	**Conceptual Subitizer with Place Value and Multiplication**. Verbally labels structured arrangements shown only briefly, using groups, multiplication, and place value. "I saw groups of 10s and 3s, so I thought, 3 tens is 30 and 4 threes is 12, so 42 in all."	*Snapshots with Dots*. Play "Snapshots" with structured groups that support the use of increasingly sophisticated mental strategies and operations. 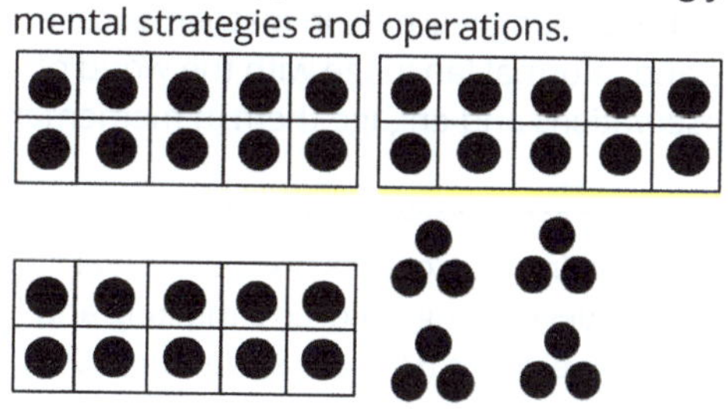

***FIGURE 4.1.* A learning trajectory for recognition of number and subitizing.**

Note. Adapted from Clements & Sarama, pp. 17–20, copyright 2014, *Learning and teaching early math: The learning trajectories approach* (2nd ed.) by Clements & Sarama. Adapted with permission of Taylor and Francis Group, LLC, a division of Informa plc.

STANDARD C.2. PEDAGOGICAL KNOWLEDGE AND PRACTICES FOR TEACHING MATHEMATICS

Well-prepared beginning teachers of mathematics have foundations of pedagogical knowledge, effective and equitable mathematics teaching practices, and positive and productive dispositions toward teaching mathematics to support students' sense making, understanding, and reasoning.

Indicators include

C.2.1. Promote Equitable Teaching
C.2.2. Plan for Effective Instruction
C.2.3. Implement Effective Instruction
C.2.4. Analyze Teaching Practice
C.2.5. Enhance Teaching Through Collaboration With Colleagues, Families, and Community Members

Teaching mathematics is complex. It entails not only knowing the mathematics but also knowing how to design and implement rich mathematics learning experiences that advance children's mathematical knowledge and proficiencies. Effective teachers are skilled in their use of high-leverage mathematics-teaching practices and use those pedagogical practices to guide both their preparation and enactment of mathematics lessons. The development of these content-focused skills and abilities (i.e., teaching practices specific to mathematics) form the core of work in the preparation of early childhood teachers of mathematics. Therefore, we have a critical elaboration of this standard focused on the knowledge and pedagogical practices specific to the early childhood level.

EC.4. Tools, Tasks, and Talk as Essential Pedagogies for Meaningful Mathematics

> Well-prepared beginning teachers of mathematics at the early childhood level intentionally plan for and use tools, tasks, and talk as pedagogies for young children's engagement in meaningful mathematics. [Elaboration of C.2.2 and C.2.3]

Effective teaching entails meeting children where they are mathematically on a learning trajectory and employing instructional pedagogies that utilize tools, tasks, and talk to support advancement in children's mathematical understanding and skills. Candidates in effective teacher preparation programs learn to ask and answer these fundamental questions: Is this (child) group of children on the learning trajectory as expected for (his or her) their ages and grades? If not, where are they on the trajectory? Where do they need to move next mathematically? How can I, as their teacher, provide instructional experiences that help them progress in their understanding and use of mathematics? What instructional tasks, tools, and activities might be most beneficial to support their learning? How can I engage these young learners in mathematical conversations and discourse that helps them connect their experiences and informal language with the world of mathematics? What questions should I ask them to draw out their observations and engage them in mathematical talk?

Well-prepared beginners select tasks purposefully and prompt children to use tools in solving mathematical problems to support children's progress on specific learning trajectories. They know that not all tasks provide the same opportunities for developing children's thinking and learning and that even young children need regular experiences with high-level tasks (Hiebert & Wearne, 1993; Stein et al., 1996). They also know that young children are good problem solvers and learn through problem-solving experiences (Cai, 2003; Moser & Carpenter, 1982). These teachers see their roles as helping children mathematize their worlds while nurturing understanding of mathematical concepts and relationships and developing language to talk about those emerging observations. They situate mathematical tasks in children's ways of knowing and learning, including attending to children's cultures, languages, genders, socioeconomic statuses, cognitive and physical abilities, funds of knowledge, and personal interests. In addition, these teachers know not to rush children toward procedural fluency; they recognize that such fluency builds from conceptual understanding through use of informal reasoning strategies in solving problems and that it develops over long periods of time, from months to years (NCTM, 2014a; NRC, 2001a).

Effective teachers learn to see and view mathematics through the eyes of their students, especially in the early childhood years when children's conceptions can be quite different from those of the teacher (see Vignette 4.2). Teachers know that all mathematical ideas are abstract, and learners have access to those ideas only through representations (NRC, 2001a). This challenge is especially prevalent among young children who are newly experiencing ways to mathematize their experiences and observations with physical objects, verbal analogies, and drawings as well as with invented and standard symbolic representations. Thus, well-prepared beginners demonstrate their own representational competence in using physical, visual, verbal, symbolic, and contextual representations appropriate for early mathematics; they recognize that these representations are foundational ideas for later mathematics. In addition, they know how to strategically use mathematical tools (e.g., part-whole mats, number bonds, Rekenreks/math racks, ten frames, bead strings, and tape diagrams) to develop and advance children's mathematical understanding and skills.

The work of teaching is complex, regardless of the age of the students. This work includes not only attention to tasks and tools for mathematical inquiries but also attention to involving young learners in meaningful mathematical talk or discourse when they engage in structured mathematical activities as well as in play (Van Oers, 2010). "Mathematical discourse includes the purposeful exchange of ideas through classroom discussion, as well as through other forms of verbal, visual, and written communication" (NCTM, 2014a, p. 29). Mathematics talk is particularly important for young children, given their limited but emerging abilities to write words and use mathematical symbols. Well-prepared beginners know that young children need many opportunities to talk about their mathematical observations and emerging ideas as well as to listen to and learn from their peers. Thus, well-prepared beginners become skillful in noticing and eliciting children's thinking and then engaging them in learner-focused dialogue, or mathematics talk, that links children's informal experiences and words to more formal mathematical ideas, using language appropriate for the particular learners (Rudd, Lambert, Satterwhite, & Zaier, 2008; Whitin & Whitin, 2000).

VIGNETTE 4.2. BUILDING FROM WHAT CHILDREN UNDERSTAND MATHEMATICALLY

The candidates in a mathematics methods course were asked to read the short narrative shown below and analyze the thinking of two children, Ethan and Morgan.

Both Ethan and Morgan are 4 years old and are enrolled in a preschool program. This is Morgan's second year in the program, whereas Ethan is new to the program this year. One day the teacher checked their subitizing abilities through an informal assessment. Working with one child at a time, she flashed each dot pattern shown below for a few seconds and then asked the child, "How many dots did you see?" If the child hesitated or was unsure, the pattern was shown again or given to the child to examine more closely. Then the child was asked to explain what he or she saw and how he or she determined the total number of dots.

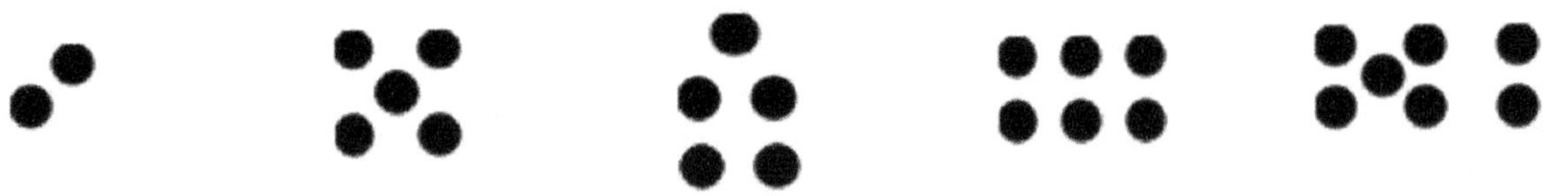

(A) (B) (C) (D) (E)

Ethan was able to identify two dots at a glance (Pattern A) but needed to count the other sets of dots by ones by touching each dot with his finger. For Pattern C, he guessed, "Eight," and for Pattern E he said, "I call that one 11." When he was handed the dot patterns, he counted the dots by ones and was able to recite the number names in the correct order, but he often had to recount the sets because he often lost track of which dots he had counted and which still needed to be counted. He demonstrated a connection between his counting and the cardinality of each set.

Morgan was able to quickly identify the total number of dots in each set without having to count by ones. For Patterns A, B, and D, the familiar dice configurations, she very quickly identified the total number of dots as two, five, and six. When asked, "How do you know it's five?" for Pattern B, she replied, "Cuz there's two and two and one in the middle. Five!" For Pattern D, she explained, "Cuz they're in the right order." She hesitated for only a moment with Pattern C and then said it was five dots and explained, "'Cuz these are four and this is five, but not in the middle" as she pointed to the familiar arrangement of four dots on the bottom of the card and then pointed to the one dot on top.

Working in pairs, the candidates placed each child's performance on the *subitizing* learning trajectory and then made instructional suggestions for advancing the learning of each child. This activity included identifying specific tasks and useful representations, formulating purposeful questions to ask the children, and articulating the mathematical knowledge and reasoning being targeted with each task. Then the whole class reconvened and collectively reached consensus on each child's learning-trajectory placement and next instructional steps. (Based on Huinker, 2011)

EC.5 Understanding Young Children's Mathematical Thinking Informs Teaching

> Well-prepared beginning teachers of mathematics at the early childhood level elicit and analyze young children's mathematical thinking to inform classroom interactions and instructional decisions. [Elaboration of C.2.3]

Educational assessments serve a variety of purposes. Sometimes assessments are high-stakes and are summative in nature. Other times, they serve a diagnostic function, as in the identification of children with special needs. Within the daily classroom interactions of the teacher and students, assessment is formative in nature; it guides instruction and gauges whether instruction has been effective or needs modification. Research and expert opinion (Clarke, Clarke, & Roche, 2011) indicate that the primary goal of assessing young children is to understand children's thinking and knowledge and to inform ongoing teaching efforts. Performance tasks (meaningful activities that require children to synthesize and apply knowledge and skills by making a response or creating a product) and informal ongoing assessments such as observations, interviews/questioning sequences, paper-and-pencil tasks, computer-based tasks, and digital records (audio and video) are useful and informative ways of assessing young children's mathematical learning and are integrated as appropriate into the early childhood mathematics curriculum.

Performing effective formative assessment supportive of early childhood learning requires that teachers understand developmental progressions of learning trajectories and set goals that are both appropriate for the children in their classrooms and responsive to each child's learning needs. They also need to know *how* to assess children to identify their strengths and current levels of understanding and what levels they are likely to learn next. Vignette 4.3 illustrates such a sequence of eliciting and using children's thinking to inform and adjust instruction. The data about children's current thinking support the team while they collaboratively determine how the current thinking connects to the progression and ways they may support movement along the content trajectory and development of the children's engagement with the mathematical practices.

VIGNETTE 4.3. COLLABORATIVE SENSE MAKING OF CHILDREN'S MATHEMATICAL THINKING

The candidates in a mathematics methods course were asked to conduct a Number Talk in their classroom setting. They engaged in preplanning the Number Talk by anticipating children's thinking and planning for how they might record their ideas using images, numbers, or both and mathematics symbols to convey the children's thinking. Next, they reviewed the plan with their cooperating teacher (or university supervisor) who made suggestions prior to implementation and took notes about the children's thinking during the lesson. After the lesson, they debriefed to consider next instructional steps on the basis of the thinking that emerged.

A teacher candidate in a first-grade classroom in February provided her children with images of dots in a tens frame. She was interested in having children conceptualize 8 in different ways. She also wanted to illustrate for students ways of recording their observations with words and number sentences. After meeting with her cooperating teacher, she also realized that this task would enable them both to see the extent to which children were subitizing. She planned to flash the image for 3 seconds, then again for another 3 seconds, and then discuss what children saw. After discussing the initial arrangement of dots (Formation 1), she rearranged the dots to see what else children may notice (Formation 2).

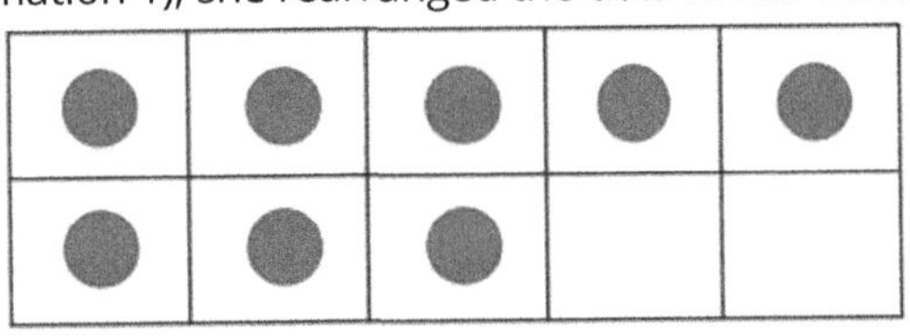

Formation 1

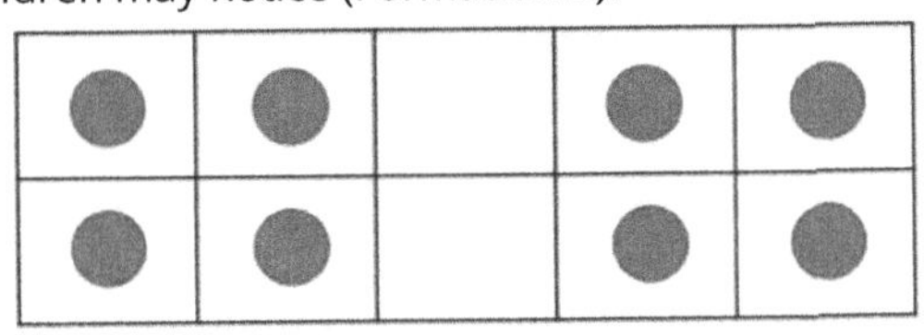

Formation 2

Some observations children made about Formation 1 were

- I noticed 5 on the top and 3 on the bottom.
- I counted really fast by 1s.
- I knew it was a 10 frame and there were 2 spots left. That's 8.
- I knew a whole row is 5 then 3 more is 8.

Observations for Formation 2 were

- I noticed 4 on one side and 4 on the other. 4 + 4 = 8.
- I saw that you didn't take any off.
- I know 2 + 2 + 2 + 2 = 8.
- It's still the same thing.

When debriefing the lesson, both the candidate and cooperating teacher were impressed with the children's thinking and participation. They decided that the Number Talk was a simple yet effective approach to gather children's ideas even though they did not hear ideas from every child. It had helped that the candidate asked the children to show a thumbs-up signal to let her know that they understood or used the same strategy as the child sharing. The cooperating teacher pointed out that the children were subitizing to 10 and connecting their counting strategies to what they know about the operations and tenness. The candidate and cooperating teacher discussed ways that the candidate could more clearly record the children's ideas (i.e., allowing more space to add connections, changing pen color with different ideas) and how together they could continue to develop the children's skills with listening and making sense of others' ideas.

The purposes of any assessment determine the content, the methods of collecting evidence, and the nature of the possible consequences for individual students, teachers, schools, or programs. Assessment that supports early childhood learning includes both formal and informal assessments and draws upon a range of sources of evidence. It enhances teachers' powers of observation and understanding of children's mathematical thinking, enabling them to establish individual learning goals. Strategies include individual assessment, observations, documentation of children's talk, interviews, samples of children's work, and performance assessments with the intent to illuminate children's thinking, strengths, and needs.

EC.6. Collaboration With Families Enhances Children's Mathematical Development

> Well-prepared beginning teachers of mathematics at the early childhood level collaborate with families in a mutually respectful, reciprocal manner to enhance and connect children's in-school and out-of-school mathematical development. [Elaboration of C.2.5]

Early childhood teachers positively influence the relationships between families and schools and the attitudes of their children toward school. They exert this influence by connecting and communicating with families in culturally sensitive ways. They develop home-school communication that allows two-way sharing of information, concerns, and feelings. They get to know families, develop trusting relationships, and collaborate on behalf of their children. They use multiple methods to communicate the mathematics their children are learning and invite families to share the ways they were taught mathematics when they were children. By inviting parents to share their ways of doing mathematics, teachers learn to communicate how their learning is similar to and different from the ways their children are learning mathematics in school. By connecting with parents, well-prepared beginners engage in the mutual sharing of resources and ideas to support the mathematical development of young learners.

Well-prepared beginners demonstrate interest in learning how their children and their families use mathematics at home and in their communities. For example, teachers can ask families to teach them the first 10 number names in their home language (e.g., uno, dos, tres, ...) and use the names when the class counts small quantities. They can also invite families to collect home artifacts for the classroom to explore mathematical concepts For example, empty boxes of commonly used household products can be used when discussing geometric shapes. These activities incorporate the child's culture in the classroom and provide cultural and learning experiences for each and every student. Collaboration with families can be enhanced when parents/caregivers and children engage in mathematics tasks together (Mistretta, 2013). *Family Mathematics Nights* or a *Family Math Saturday* can provide opportunities for children to share what they are learning in mathematics and for families to engage in problem-solving activities with their children. For these activities, beginners invite parents to share their strategies for solving problems while they become more aware of their child's mathematical thinking and learning. Well-prepared beginners forge partnerships with families and support them with knowing how to create and sustain learning opportunities at home. They may also support joint and separate sessions for parents and children at the school, including bridging activities for parents to develop their child's numeracy at home (Doig, McCrae, & Rowe, 2003).

STANDARD C.3. STUDENTS AS LEARNERS OF MATHEMATICS

Well-prepared beginning teachers of mathematics have foundational understandings of students' mathematical knowledge, skills, and dispositions. They also know how these understandings can contribute to effective teaching and are committed to expanding and deepening their knowledge of students as learners of mathematics.

Indicators include

C.3.1. Anticipate and Attend to Students' Thinking About Mathematics Content

C.3.2. Understand and Recognize Students' Engagement in Mathematical Practices

C.3.3. Anticipate and Attend to Students' Mathematical Dispositions

Effective teachers understand how students' mathematical ideas develop and how to apply such understandings to every aspect of teaching. Such understanding is particularly important at the early childhood level because children often interpret mathematical situations, even those that seem obvious to adults, quite differently from adults. Therefore, we have a critical elaboration of this standard for early childhood mathematics education.

EC.7. Seeing Mathematics Through Children's Eyes

> Well-prepared beginning teachers of mathematics at the early childhood level are conversant in the developmental progressions that are the core components of learning trajectories and strive to see mathematical situations through children's eyes. [Elaboration of C.3.1]

The younger the child the more important teachers' uses of children's thinking and learning are as starting points. Further, the younger the child, the more difficult *decentering* and seeing mathematical situations through children's eyes will be. Fortunately, we know in great detail how children think and learn mathematics in the early years, and for candidates to become conversant in these *developmental progressions* and utilize them is important when they plan for and interact with children in preschool and primary settings.

These developmental progressions are paths most children follow in learning a mathematics topic. These paths are children's natural ways of learning. For example, consider that children first learn to crawl, then walk, then run, skip, and jump with increasing speed and dexterity. These are the levels in that developmental progression of movement. Children similarly follow natural developmental progressions in learning the concepts and skills within a certain domain or topic of mathematics. When teachers understand these developmental progressions, and select, sequence, and modify activities on the bases of them, they create mathematics learning environments that are particularly developmentally appropriate and effective. These developmental progressions, then, are the core of a learning trajectory (which includes, as described previously, the mathematical goal, or content, and instructional activities and strategies corresponding to each level of the developmental progression).

Developmental progressions begin when life begins. Young children have certain mathematical-like competencies in number, spatial sense, and patterns from birth. While they develop and learn, they progress through identifiable *levels of thinking*—periods of time of qualitatively distinct patterns of thinking about mathematics. As an example, children develop increasingly sophisticated counting strategies to solve increasingly difficult types of arithmetic problems. For example, even very young children, shown one chip on a plate covered then shown another chip placed under the cover, can make their plates "look just like mine"—an early *visual addition*. Later, these children use a counting-all procedure. Given a situation such as combining six red apples and two green apples, children count out objects to form a set of six items, then count out two more items, and finally count all those items and say, "Eight." After children develop such methods, they eventually phase them out, in favor of other methods. On their own, children as young as 4 or 5 years may start *counting on,* solving the previous problem by counting, "Siiiix, ... seven, eight. Eight!" The elongated pronunciation of the first addend substitutes for counting the initial set one-by-one. That approach is used as if they first counted a set of six items.

Thus, counting skills—especially sophisticated counting skills—play an important role in developing competence with computation. Counting-on when increasing collections and the corresponding *counting-back-from* when decreasing collections are critical numerical strategies for children. However, they are only beginning strategies. If the amount of increase is unknown, children use *counting-up* to find the unknown amount. If five items are added to so that one now has nine items, children may find the amount of increase by counting and keeping track of the number of counts, as in (with drawn-out pronunciation of "five" and later, "nine"): "Fiiiive; 6, 7, 8, 9. Four (as in four counts, the amount of increase)!" And if items are removed from nine items so that five remain, children may count back from nine to five to find the unknown decrease as follows: "Niiiine; 8, 7, 6, 5, 4. Four!!" However, counting backward, especially more than two or three counts, is difficult for most children unless they have consistent instruction. Instead, children might learn counting-up to the total to solve a subtraction situation. For example, "I took away 5 from 9, so 6, 7, 8, 9 (raising a finger with each count)— that's 4 more left in the 9." Learning this way, including the complementary use of number composition and decomposition strategies (e.g., break apart to make a 10) is more developmentally appropriate and more effective at achieving fluency than jumping immediately to verbal memorization (Baroody, 1999; Henry & Brown, 2008).

STANDARD C.4. SOCIAL CONTEXTS OF MATHEMATICS TEACHING AND LEARNING

Well-prepared beginning teachers of mathematics realize that the social, historical, and institutional contexts of mathematics affect teaching and learning and know about and are committed to their critical roles as advocates for each and every student.

Indicators include

C.4.1. Provide Access and Advancement
C.4.2. Cultivate Positive Mathematical Identities
C.4.3. Draw on Students' Mathematical Strengths
C.4.4. Understand Power and Privilege in the History of Mathematics Education
C.4.5. Enact Ethical Practice for Advocacy

Effective teachers are attuned to the specific strengths and backgrounds of each of their students. They build on a student's current mathematical ideas and ways of knowing and learning, including attending to the student's culture, race/ethnicity, language, gender, socioeconomic status, cognitive and physical abilities, and personal interests. They also attend to developing positive mathematical identities and agency among their students. Such understanding is particularly important at the early childhood level, when children take their first steps from their home lives to the world of formal education. Therefore, we have a critical elaboration of this standard focused on the knowledge and contexts specific to the early childhood level.

EC.8. Creating Positive Early Childhood Learning Environments

> Well-prepared beginning teachers of mathematics at the early childhood level create mathematical learning environments characterized by exploration, reasoning, and problem solving; they draw upon children's mathematical, cultural, and linguistic strengths thereby developing conceptual understanding and positive mathematical identities. [Elaboration of C.4.2 and C.4.3]

Classroom learning environments provide contexts to shape the ways children experience and learn mathematics. From the visual displays on the classroom walls to the arrangement of desks and the accessibility of instructional materials, the learning environment indicates how a teacher views mathematics and what children are expected to learn and do. Well-prepared beginners understand that learning environments affect young learners' developing mathematical identities and can either support or hinder children's abilities to learn mathematics. By using classroom routines as opportunities to explore mathematics, these teachers create learning opportunities in which all children feel invited to participate. For example, when kindergartners are lined up for lunch, a teacher might facilitate an exploration of the various meanings of counting, such as ordinal numbers (e.g., first, second, third in line). Similarly, a teacher may further develop first graders' understandings of measurement ideas by asking the children to compare the heights of those in line and order them from tallest to shortest. Second graders may be asked to examine patterns and concepts of odd and even when they consider whether every child has a partner when they form two lines. Whether children are counting out snacks to place in baggies or dividing classroom supplies for a mathematical task, well-prepared beginners learn to attend to these events to deepen young children's learning of mathematical concepts.

Well-prepared beginners understand the role of manipulative materials in helping young learners represent mathematical concepts and communicate their thinking. They select and use a variety of manipulative materials and encourage children to use these tools to explore mathematical ideas without always telling them when and how to use the materials. Materials are easily accessible for young learners to use when they explore mathematical concepts, solve problems, and communicate their mathematical thinking for themselves and others. Well-prepared beginners also know that the type of manipulative materials and the names they give to them support children's later learning of mathematics. They use manipulatives to help young children make connections among concrete counters, number names, and symbols when children transition from play-based, real-world knowledge of quantities to formal base-ten number systems (Morin & Samelson, 2015).

Well-prepared beginners establish classroom norms that reflect their valuing of the various ways children explore and reason about mathematical situations and support young learners' development of mathematical explanations (Yackel & Cobb, 1996). They know that language and communication patterns are functions of a person's culture and understand that children enter classrooms speaking the language and language vernaculars used by their families and friends outside of school. These teachers treat children's language as a resource, not a deficit, and support all children, regardless of their English proficiency, to participate in class discussions (Moschkovich, 2010). Well-prepared beginners see their roles as helping young learners connect their out-of-school communication practices with the academic and mathematical language they are expected to use in schools. They engage their children in classroom discussions and encourage children to use their existing communication patterns while they teach them the language of mathematics, including vocabulary, symbols, and materials. Consider Vignette 4.4.

VIGNETTE 4.4. SOLVING 21 + 32

A class of first graders is finding the total number of stickers in two different-sized packages. One package has 21 stickers, and the other package has 32 stickers. The children can use connecting cubes, base-ten blocks, or paper and pencil to solve the problems. After the children solved the problem, the teacher, Mr. Walker, engages the class in a discussion.

Teacher: Who would like to share how they solved the problem? Rebecca?

Rebecca: I used the sticks and cubes [Points to base-ten blocks]. I got two sticks and one cube and three sticks and two cubes and put them together. Then I got 53.

Teacher: How did you know to take out 2 tens and 1 unit to equal the number 21?

Rebecca: Cause the stick is 10.

Teacher: The stick is 10 what?

Rebecca: Ten tinies. So one stick is 10, and two sticks are 20, and I needed one more to equal 21.

Teacher: Will you count the tens and ones for us?

Rebecca: [Touches each block when she counts] Ten, 20, 30, 40, 50, 51, 52, 53.

Teacher: Thank you; I see what you did now. Did anyone solve the problem a different way? Jamaal?

Jamaal: I did it in my head. I plussed 20 and 30 and got fiddy. Then I plussed the one and the two and got three. So my answer is fiddy-three.

Teacher: [Writes 53 on the board]. So let me see if I understand what you did. First you added 20 and 30 and got the sum of 50. Then you added one and two and got the sum of three. Then what did you do?

Jamaal: I plussed, I mean added, *fiddy* and three and got fiddy-three.

Teacher: I see, thank you for sharing.

In this exchange, Mr. Walker builds on Rebecca's and Jamaal's everyday language and provides them with the mathematical terms to use when they communicate. Rather than correcting children when they speak, he revoices their statements using mathematical terms. More important, in the case of Jamaal, Mr. Walker realizes that he said "fiddy" instead of "fifty" but does not correct his language because he understands what Jamaal means and does not want him to feel uncomfortable in sharing his thinking. Instead Mr. Walker uses the correct mathematical pronunciation and continues the class discussion. Well-prepared beginners build on children's mathematical, cultural, and linguistic strengths to promote the positive and mathematical dispositions of every child.

PART 2. ELABORATIONS OF THE CHARACTERISTICS NEEDED BY EFFECTIVE PROGRAMS PREPARING EARLY CHILDHOOD TEACHERS OF MATHEMATICS

This section provides additional detail of what preservice programs need to do to effectively prepare candidates to teach early childhood mathematics. Although young children are ready and eager to learn (NRC, 2001b), many early childhood teachers are not eager and prepared to engage children in rich experiences in domains other than literacy (Brenneman, Stevenson-Boyd, & Frede, 2009; IOM & NRC, 2015; NRC, 2007). Teachers of young children historically have not been prepared to teach domain-specific knowledge to young children (Isenberg, 2000). Effective programs ensure that they follow all the recommendations in Chapter 3 to provide guidance in enacting research-based policies and practices for the preparation of all early childhood teachers of mathematics. Here we provide brief elaboration of three of the program standards, along with commentary on the other two program standards.

STANDARD P.1. PARTNERSHIPS

An effective mathematics teacher preparation program has significant input and participation from all appropriate stakeholders.

Indicators include

P.1.1. Engage All Partners Productively
P.1.2. Provide Institutional Support

Effective programs preparing teachers of mathematics at the early childhood level include collaboration among the wide range of stakeholders, such as mathematics teacher educators and researchers, their teacher preparation colleagues, preschool and primary grade educators, mathematicians, and community members. Their multiple perspectives and knowledge bases ensure that programs will effectively deepen candidates' mathematical knowledge relevant to their teaching and develop mathematical pedagogies and effective teaching practices as well as address the diverse needs of its communities.

Strong partnerships engage all partners in developing a common vision and identifiable goals: the preparation of high-quality early childhood teachers of mathematics. The partnership ideally includes a range of stakeholders, at the minimum, mathematics teacher educators and researchers, their teacher preparation colleagues, preschool and primary grade educators, mathematicians, and community members (these stakeholders and more are essential, and all professionals need scientific knowledge of young children's development; see IOM & NRC, 2015). Teacher preparation programs benefit from collaborative partnerships that bring together multiple perspectives to ensure that programs deepen candidates' mathematical knowledge relevant to their teaching and develop mathematical pedagogies and effective teaching practices as well as address the diverse needs of its communities. Refer to Chapter 3 for specific guidance on building effective partnerships to support the preparation of well-prepared beginning early childhood teachers of mathematics.

STANDARD P.2. OPPORTUNITIES TO LEARN MATHEMATICS

An effective mathematics teacher preparation program provides candidates with opportunities to learn mathematics and statistics that are purposefully focused on essential big ideas across content and processes that foster a coherent understanding of mathematics for teaching.

Indicators include

P.2.1. Attend to Mathematics Content Relevant to Teaching
P.2.2. Build Mathematical Practices and Processes
P.2.3. Provide Sustained Quality Experiences

Effective programs preparing teachers of mathematics at the early childhood level include at least one and preferably more mathematics content courses or other high-quality equivalent professional-learning experiences focused on key mathematics concepts, skills, and processes important to the early grades. Such experiences provide opportunities for teacher candidates to develop in-depth understanding of the surprisingly complex ideas that form the foundation of all later mathematics.

EC.9. Mathematics Content Preparation of Early Childhood Teachers

> Effective programs preparing teachers of mathematics at the early childhood level require at least one mathematics content course (or equivalent professional-learning experiences) focused on key mathematical ideas and processes that are important in early mathematics, including problem solving, number, operations, spatial thinking, shapes, measurement, and early algebraic thinking. [Elaboration of P.2]

Early childhood educators are responsible for establishing the mathematical foundation for children from birth to age 8 years. This is an especially critical period for laying the groundwork of mathematics ideas and understandings as well as forming early positive attitudes toward mathematics and statistics. Well-prepared beginners need to study the mathematics content relevant to children's learning of mathematics in preschool and the primary grades as well as the mathematics that extends through the elementary grades. Understanding this range and progression of major mathematical domains enables well-prepared beginners to support programmatic coherence between preschool and the primary grades and from the primary grades into upper elementary school. For well-prepared beginners to utilize the learning trajectories, they must understand the underlying mathematics—a core element of learning trajectories (too often neglected). The trajectories highlight mathematical content and proficiencies important for early childhood educators, including the importance of both composing and decomposing number, composition and decomposition of shapes, spatial thinking, subitizing, angle size, length, area, and volume as well as early multiplication and division ideas and early fractions.

For these reasons, effective programs for preparing teachers of mathematics at the early childhood level and others who support young children's learning of mathematics provide opportunities that enable candidates to learn mathematics content and practices throughout the range of the preschool and elementary school. Tables 4.2, 4.3, and 4.4, highlight the important mathematical concepts for well-prepared beginners to know and understand deeply. These concepts, identified in *The Mathematical Education of Teachers II* (CBMS, 2012), focus on counting and cardinality, operations and algebraic thinking, number and operations in base ten, measurement and data, and geometry. Further, programs immerse candidates in such mathematical practices as reasoning, sense making, and problem solving while they learn content. Candidates learn to explain their thinking, recognize structures, and generalize. Program personnel guide candidates in making mathematical connections among approaches to solving problems, among mathematical topics (e.g., measurement of area and multiplication), and between mathematics and other disciplines. Programs employ teaching strategies consistent with those that are effective with young children. Such programs develop positive dispositions toward mathematics, including persistence and a desire to engage in posing and solving problems.

Studying mathematics content is necessary but not sufficient. High-quality early childhood teacher preparation programs weave together the learning of mathematics content, the study of specific mathematics pedagogies and effective mathematics instruction, and, at the core, developmental knowledge of children's mathematical thinking and reasoning.

STANDARD P.3. OPPORTUNITIES TO LEARN TO TEACH MATHEMATICS

An effective mathematics teacher preparation program provides candidates with multiple opportunities to learn to teach through mathematics-specific methods courses (or equivalent professional learning experiences) in which mathematics, practices for teaching mathematics, knowledge of students as learners, and the social contexts of mathematics teaching and learning are integrated.

Indicators include

P.3.1. Address Deep and Meaningful Mathematics Content Knowledge
P.3.2. Provide Foundations of Knowledge About Students as Mathematics Learners
P.3.3. Address the Social Contexts of Teaching and Learning
P.3.4. Incorporate Practice-Based Experiences
P.3.5. Provide Effective Mathematics Methods Instructors

EC.10. Mathematics Methods Experiences for Early Childhood Teachers

> Effective preparation programs preparing teachers of mathematics at the early childhood level require at least one mathematics methods course (equivalent of 3 semester units) focused on mathematics teaching, children's mathematical thinking, and development of mathematics learning at the early childhood level. [Elaboration of P.2 and P.3]

The demands of teaching and learning of mathematics in the 21st century require that candidates complete at least one mathematics methods course focused on grades pre-K through Grade 2 to fully prepare new teachers with a strong foundation for success. In no grade-bands must more work be done to achieve this than in pre-K–2. Given the goal of educating mathematics teachers who are well-prepared, the often challenging nature of initial teaching placements, and the breadth and depth of practices and dispositions needed, programs designed to prepare beginning mathematics teachers provide multiple mathematics methods courses for those seeking certification to teach broader grade-bands (such as Pre-K–5 or K–8). At the same time we acknowledge that differences in the designs of teacher preparation programs will shape the ways in which this goal is achieved. As a field we need to subject these approaches to appropriate scrutiny to ensure that programs graduate early childhood teachers of mathematics who are well prepared.

Methods courses include opportunities for teacher candidates to engage in mathematics, learn about children's mathematical thinking and solution strategies, identify and integrate children's lived experiences and funds of knowledge into mathematics lessons, learn pedagogical strategies discussed in Chapter 2, and learn to use assessment to build on children's understandings and support and extend their learning—in short, to understand and use complete learning trajectories in their teaching.

STANDARD P.4. OPPORTUNITIES TO LEARN IN CLINICAL SETTINGS

An effective mathematics teacher preparation program includes clinical experiences that are guided on the basis of a shared vision of high-quality mathematics instruction and have sufficient support structures and personnel to provide coherent, developmentally appropriate opportunities for candidates to teach and to learn from their own teaching and the teaching of others.

Indicators include

P.4.1. Collaboratively Develop and Enact Clinical Experiences
P.4.2. Sequence School-Based Experiences
P.4.3. Provide Teaching Experiences With Diverse Learners
P.4.4. Recruit and Support Qualified Mentor Teachers and Supervisors

Effective preparation programs of beginning early childhood teachers of mathematics include clinical experiences, in both preschool and primary settings, that are exemplar sites in equitable early mathematics learning. These sites illustrate responsive interactions with children individually, in small groups, and as a class.

EC.11. Clinical Experiences in Mathematics for Early Childhood Teachers

> Effective programs preparing teachers of mathematics at the early childhood level provide clinical experiences specific to mathematics focused on children's mathematical thinking and mathematics instruction with diverse learners in preschool and primary settings. [Elaboration of P.1 and P.4]

Experiences in both preschool and primary settings include "field observations, field work, practica, student teaching and other 'clinical' practice experiences such as home visiting" (NAEYC, 2012, Initial and Advanced Standards). These experiences provide opportunities for systematic inquiry into classroom practice under the supervision of licensed professionals with the intention of preparing teachers of young children who develop nurturing, responsive relationships with children and families. Additional recommendations can be found in Standard 3.4. Opportunities to Learn in Clinical Settings.

Effective programs benefit from developing partner schools that host groups of teacher candidates, pairing them with cooperating teachers who are willing to share their classroom practices in ways that benefit learning for the young learners, the teacher candidates, and the cooperating teachers themselves. Such a setting provides a space for close examination of instructional practice as highlighted in Vignette 4.3. Teacher candidate/cooperating teacher teams collaboratively plan, teach, and debrief lessons on the basis of student data (Rigelman & Ruben, 2012). When purposefully designed, the clinical experiences can be explicitly linked to the mathematics methods coursework to narrow the theory-practice gap lamented by many teacher educators (Darling-Hammond & Bransford, 2007). These strong partnerships provide space for preservice and in-service teacher learning in support of improved mathematics learning for children.

STANDARD P.5. RECRUITMENT AND RETENTION OF TEACHER CANDIDATES

An effective mathematics teacher preparation program attracts, nurtures, and graduates high-quality teachers of mathematics who are representative of diverse communities.

Indicators include

P.5.1. Recruit Strong Candidates
P.5.2. Address Diverse Community Needs
P.5.3. Provide Experiences and Support Structures

Our youngest children deserve teachers who feel passion for mathematics. In recruiting high-quality teacher candidates at the early childhood level, program personnel seek out individuals who enjoy the study of mathematics and look forward to the teaching of mathematics. Admission requirements include multiple measures of not only cognitive factors (e.g., mathematics course grades, mathematics test scores) but also dispositional factors. Through interviews or essays, personnel can identify individuals eager to learn effective methods for teaching mathematics as well as uncover implicit biases or deficit views of diverse children and families and such dispositions to avoid mathematics as "I was never good at math" or "I'm just not a math person." Applicants to early childhood education programs often view themselves with strengths in literacy and weaknesses in mathematics. Preparation programs can change these views by disseminating contrasting views of the joy and intellectual stimulation that early mathematics education provides to both teachers and children and by developing candidates' knowledge and skills in mathematics and mathematics education. Programs can use both cognitive and dispositional factors related to mathematics to inform both admissions and program planning decisions, for example, identifying and building supports that contribute to the short- and long-term successes of their teacher candidates in teaching mathematics to young children.

Teacher retention is a challenge, especially retaining teachers of the youngest children. As described in Chapter 3, successful recruitment and retention strategies include field experiences that provide positive experiences with high-quality mathematics education in both preschool and the primary grades, scholarships specific to teaching mathematics to young children, clear foci on the integrated and active-learning approach so effective with young children, and career counseling in early education. Teacher education programs monitor the early career progress of their graduates to inform program improvement as well as to contribute to the retention of teachers.

CLOSING REMARKS

Effective programs for preparing teachers of mathematics at the early childhood level develop candidates' abilities to use high-leverage, effective mathematics teaching practices (NCTM, 2014a) that require deep understandings of the mathematics candidates are expected to teach. The teaching they are expected to enact on a daily basis with young learners often stands in sharp contrast to what many candidates experienced themselves as learners of mathematics (Isenberg, 2000). They often describe their own experiences as being teacher-centered instruction emphasizing memorization of facts and procedures with little to no emphasis on understanding, problem solving, reasoning, and application. In addition, many early childhood teachers report high levels of mathematics anxiety and avoidance. Effective programs consciously break the insidious cycle that currently exists in which early childhood teachers pass their own anxieties and superficial knowledge of mathematics on to their children. Instead, effective programs prepare their candidates to engage in *ambitious teaching*, teaching in which teachers "aim to teach all kinds of children to not only know academic subjects, but also to be able to use what they know in working on authentic problems in academic domains" (Lampert, Boerst, & Graziani, 2011, p. 1). In addition, effective programs instill in their candidates positive and productive dispositions toward mathematics teaching and learning, which they will, in turn, pass on to their students.

CHAPTER 5. ELABORATIONS OF THE STANDARDS FOR THE PREPARATION OF UPPER ELEMENTARY GRADES TEACHERS OF MATHEMATICS

Teachers who teach mathematics and statistics in upper elementary grades must have not only strong general teaching skills but also strong content knowledge, strong knowledge of mathematics-specific pedagogy, and much more, including cultural knowledge about their individual students, school policies, and how to collaborate with other teachers. Only with this knowledge will teachers of mathematics be able to meaningfully support the learning of each and every student.

Building on the standards presented in Chapters 2 and 3, we put forth in this chapter particular elaborations of the standards for the knowledge, skills, dispositions, and actions of well-prepared beginning upper elementary grades teachers of mathematics as well as requirements for preservice programs to ensure the effective preparation of candidates to develop those necessary knowledge, skills, dispositions, and actions. Additionally, the chapter includes commentary and examples about those standards relevant to upper elementary grades. The chapter concludes with standards regarding how programs may achieve these relevant standards. The elaborations in this chapter focus on those standards with specific upper elementary considerations; therefore, although all the standards in Chapters 2 and 3 apply to upper elementary mathematics teacher candidates, not all require elaboration.

Table 5.1 lists the elaborations of the standards presented in Chapters 2 and 3 as they relate to preparing teachers of mathematics for upper elementary grades.

TABLE 5.1. ELABORATIONS OF CANDIDATE AND PROGRAM STANDARDS FOR TEACHERS OF MATHEMATICS FOR UPPER ELEMENTARY GRADES

Part 1. Candidate Knowledge, Skills, and Dispositions	
UE.1. Mathematics Concepts and Connections to Mathematical Practices	Well-prepared beginning teachers of mathematics at the upper elementary level understand foundational mathematics concepts that they will teach, and they connect those concepts to mathematical practices as well as to the mathematics of Pre-K–2 and the middle level curriculum. [Elaboration of C.1.1 and C.1.2]
UE.2. Pedagogical Knowledge and Teaching Practices	Well-prepared beginning teachers of mathematics at the upper elementary level develop pedagogical knowledge and practices to cultivate students' mathematical proficiency, including such components as conceptual understanding, procedural fluency, problem-solving ability, and facility with the mathematical processes essential for learning. [Elaboration of C.2.2 and C.2.3]
UE.3. Tools to Build Student Understanding	Well-prepared beginning teachers of mathematics at the upper elementary level effectively use technology tools, physical models, and mathematical representations to build student understanding of the topics at these grade levels. [Elaboration of C.1.6 and C.2.3]
UE.4. Assessment to Promote Learning and Improve Instruction	Well-prepared beginning teachers of mathematics at the upper elementary level learn to use both formal and informal assessment tools and strategies to gather evidence of students' mathematical thinking in ways appropriate for young learners, such as the use of observations, interviews, questioning, paper-and-pencil and computer-based tasks, and digital records, including audio and video. [Elaboration of C.3.1]
UE.5. Students' Sense Making	Well-prepared beginning teachers of mathematics at the upper elementary level nurture students' proficiency with, and making sense of, mathematical ideas, processes, and practices. [Elaboration of C.3.1 and C.3.2]
UE.6. Ethical Advocates for Students	Well-prepared beginning teachers of mathematics at the upper elementary level understand their roles as ethical advocates for elementary-grades students to have access to and advance in mathematics that cultivates positive mathematics identities and connects to students' mathematical thinking and lived experiences; these teachers build partnerships with families and communities and work to eliminate institutional and curricular barriers to learning. [Elaboration of C.4.1]
Part 2. Program Characteristics	
UE.7. Mathematical Content Preparation of Upper Elementary Grades Teachers of Mathematics	Effective programs preparing teachers of mathematics at the upper elementary level include coursework and other experiences focused on key mathematical ideas and skills that are pivotal in those grades. [Elaboration of P.2.1]
UE.8. Mathematics Methods Coursework for Upper Elementary Grades Teachers of Mathematics	Effective programs preparing teachers of mathematics at the upper elementary level include at least one mathematics methods course, or the equivalent experience of 3 semester units, focused particularly on mathematics teaching and learning in upper elementary grades. [Elaboration of P.3.1, P.3.2, P.3.3, and P.3.4]
UE.9. Clinical Experiences for Upper Elementary Grades Teachers of Mathematics	Effective programs preparing teachers of mathematics at the upper elementary level develop and systematically use a collection of clinical settings that support beginning teachers' work with diverse learners and curricula and within differing institutional contexts. [Elaboration of P.4.2 and P.4.3]

PART 1. ELABORATIONS OF THE KNOWLEDGE, SKILLS, AND DISPOSITIONS NEEDED BY WELL-PREPARED BEGINNING TEACHERS OF MATHEMATICS IN THE UPPER ELEMENTARY GRADES

This section provides additional detail, commentary, and examples of the knowledge, skills, and dispositions well-prepared beginning teachers who teach mathematics in upper elementary grades have, organized by the general standards described in Chapter 2. Standards specific to teaching mathematics in upper elementary grades are defined and described, along with additional commentary and examples about the general standards in Chapter 2. The elaborations in this chapter focus on those standards with specific considerations for upper elementary grades. Although all the standards in Chapters 2 and 3 apply to mathematics-teacher candidates for upper elementary grades, not all require elaboration.

STANDARD C.1. MATHEMATICS CONCEPTS, PRACTICES, AND CURRICULUM

Well-prepared beginning teachers of mathematics possess robust knowledge of mathematical and statistical concepts that underlie what they encounter in teaching. They engage in appropriate mathematical and statistical practices and support their students in doing the same. They can read, analyze, and discuss curriculum, assessment, and standards documents as well as students' mathematical productions.

Indicators include

C.1.1. Know Relevant Mathematical Content
C.1.2. Demonstrate Mathematical Practices and Processes
C.1.3. Exhibit Productive Mathematical Dispositions
C.1.4. Analyze the Mathematical Content of Curriculum
C.1.5. Analyze Mathematical Thinking
C.1.6. Use Mathematical Tools and Technology

Effective teachers have deep understandings of the mathematics they are expected to teach and exhibit positive dispositions toward both mathematics teaching and learning. Such understandings and dispositions are particularly critical for upper elementary teachers because they develop the foundations of mathematical understanding, beliefs, and attitudes among young learners to start students on their mathematical journeys. Therefore, we have a critical elaboration of this standard for the preparation of teachers of mathematics in upper elementary grades.

UE.1. Mathematics Concepts and Connections to Mathematical Practices

> Well-prepared beginning teachers of mathematics at the upper elementary level understand foundational mathematics concepts that they will teach, and they connect those concepts to mathematical practices as well as to the mathematics of Pre-K–2 and the middle level curriculum. [Elaboration of C.1.1 and C.1.2]

Prospective upper elementary grade teachers often enter their preservice teacher education courses with fragile mathematics identities. Many of them voice concerns about teaching mathematics because of their own, often negative, mathematics instructional experiences in Pre-K–16 education. Or, they may feel a sense of relief because they believe that learning elementary mathematics means knowing how to compute with the basic operations. They may think that because they know how to multiply, they can teach multiplication. Within elementary mathematics concepts are depth and complexity that candidates can and must understand. Even though the standard algorithms often continue to be emphasized in elementary schools, candidates may not know that the familiar algorithms that they learned in school and often used without understanding can and often are taught by linking these procedures to important mathematical structures and properties. Thus, developing the foundations of a robust mathematical-knowledge base is essential for learning how to effectively teach elementary mathematics.

Even well-prepared beginners do not learn all the mathematics content they need to teach all elementary grade levels in their preservice teacher education programs. However, they must study some key areas, and they should have opportunities to study some of these in depth. Key areas for upper elementary candidates include base-ten numbers, multiplicative structures, fractions and decimals, algebraic thinking, measurement, and geometry.

In brief summaries below, we describe the significant concepts that well-prepared beginning teachers of mathematics must know to be able to support learners in upper elementary grades and how those concepts connect with mathematical practices. The summaries include specific, but not exhaustive, examples of what is involved in understanding and being ready to teach this content. These ideas connect to and reflect the *MET II* content expectations (CBMS, 2012) and, therefore, we include the related *MET II* content expectations in each section in addition to content identified in research in mathematics teacher education. This section was also influenced by representations of core understandings found in documents such as NCTM's *Developing Essential Understandings* series (cf. Clark et al., 2010), *Curriculum Focal Points* (Schielack, et al., 2006), and the *Common Core State Standards – Mathematics* (NGA & CCSSO, 2010).

Multiplicative Structures. Well-prepared beginning teachers of mathematics in the upper elementary grades are multiplicative thinkers, with deep understandings of the following concepts and topics:

- Multiplication and division have meaning, including several interpretations, and involve more than just memorization of basic facts and procedures. Sometimes teachers (and students) believe that multiplication is merely repeated addition or equal groups and that division is only making groups, but well-prepared beginning teachers must be familiar, beyond those partial understandings, with the multiple mathematical meanings of and real-world contexts for multiplication and division. Well-prepared beginning teachers can represent these operations in many ways and can make connections between representations and problem types.
- Properties, such as the commutative, associative, and distributive properties, support justification, flexibility, and fluency and help students make sense of multi-digit computation.
- Computation involving multiplication and division includes mental computation, estimation strategies, invented algorithms, and standard algorithms (Otto, Caldwell, Lubinski, & Hancock, 2011).

Well-prepared beginning teachers of the upper elementary grades must understand the concepts of multiplication and division, including multiple representations of the concepts as well as how to sequence and teach this content to students. Teachers recognize the relationships between the content that precedes multiplication and division (e.g., addition, subtraction, and place value) and the content that follows multiplication and division (e.g., ratios and proportions). The movement from additive thinking, particularly counting by ones, to multiplicative thinking or seeing an equal group as a unit is challenging for students, and the well-prepared beginning teacher predicts and responds to that potential barrier.

Well-prepared beginners know that the way to approach this content is through engaging students in reasoning about situations involving multiplication and division. Such activities present opportunities to connect the action described in the problem with the arithmetical operations, which is also essentially the definition of the mathematical practice of *making sense of problem situations*. Well-prepared beginners know how to select and sequence problems to introduce students to a range of interpretations of these concepts and to the need to develop strategies for solving these problems through multiple approaches. These approaches include skip counting, equal groups, area and array models, multiplicative comparisons, ratio tables, scaling, and partitive- and measurement-division situations (Otto, Caldwell, Lubinski, & Hancock, 2011). Additionally, well-prepared beginners are able to select rich tasks that lead students to the use of multiple approaches, connect to relevant contexts, and connect to other content within mathematics.

Table 5.2 lists *MET II* report (CBMS, 2012) items related to multiplicative structures in the upper elementary grades.

TABLE 5.2. CONNECTIONS TO *MET II* (CBMS, 2012) RELATED TO MULTIPLICATIVE STRUCTURES IN THE UPPER ELEMENTARY GRADES

MET II describes the following essential ideas related to multiplicative structures:

- "The different types of problems solved by multiplication and division, and meanings of the operations illustrated by these problem types.
- Teaching–learning paths for single-digit multiplication and associated division, including the use of properties of operations (i.e., the field axioms)." (p. 26)
- "Recognizing that addition, subtraction, multiplication, and division problem types and associated meanings for the operations (e.g., *CCSS,* pp. 88–89) extend from whole numbers to fractions." (p. 28)

Fractions and Decimals. Well-prepared beginning teachers of mathematics at the upper elementary level have strong understandings of fractions and decimals, including the following concepts and topics:

- Fractions have multiple interpretations, including part-whole relationships, measures, quotients, ratios, and operators.
- The unit is a foundational concept, "fundamental to the interpretation of rational numbers" (Otto et al., 2011, p. 8).
- Equivalence is a key concept. Fractions can be expressed in an infinite number of equivalent fractions and in decimal form.
- Understanding the magnitudes of fractions and decimals enables students to compare and order these numbers and perform computational estimation.
- Computation with fractions and decimals builds on understanding of whole number operations, but some interpretations and contexts make more sense than others. Estimation and mental arithmetic continue to be important.

Students in upper elementary grades experience shifts in their mathematical thinking that include

- Shifting from discrete, countable quantities only to include continuous quantities.
- Shifting from one model to a variety of representations, flexibly thinking about the unit.
- Shifting from whole-number-based comparisons to equivalence-based comparisons
- Shifting from rules to making sense about operations on fractions (Otto et al., 2011).

When students experience the natural disequilibrium the occurs when making these shifts, well-prepared beginners are ready to support sense making and understanding. This is a context in which an explicit emphasis on mathematical practices can play a key role. Well-prepared beginners encourage students to communicate their reasoning, critique the reasoning of others, and develop arguments through discourse and mathematical writing. Beginning at the third grade, students will learn that an argument is a carefully crafted sequence of statements and that reasoning strategies are presented with an objective to convincing others that a claim is true or false, rather than being merely a written listing of procedural steps taken in a process or algorithm used to arrive at an answer. Students in upper elementary grades are formally introduced to mathematical argument, including a claim, justification of the claim with evidence, and warrants that connect the reasoning and evidence to the claim (Casa et al., 2016). This introduction helps set the foundation for mathematical arguments in later grades, including inductive and deductive proof and the analysis, representing, reasoning, revising, and reporting demands of mathematical modeling at the high school level.

Table 5.3 lists *MET II* report (CBMS, 2012) items related to fractions and decimals in the upper elementary grades.

TABLE 5.3. CONNECTIONS TO *MET II* (CBMS, 2012) RELATED TO FRACTIONS AND DECIMALS IN THE UPPER ELEMENTARY GRADES

MET II describes the following essential ideas related to fractions and decimals:

- "Understanding fractions as numbers that can be represented with lengths and on number lines. Using the *CCSS* development of fractions to define fractions $\frac{a}{b}$ as *a* parts, each of size $\frac{1}{b}$. Attending closely to the whole (referent unit) while solving problems and explaining solutions.
- Recognizing that addition, subtraction, multiplication, and division problem types and associated meanings for the operations (e.g., *CCSS*, pp. 88–89) extend from whole numbers to fractions.
- Explaining the rationale behind equivalent fractions and procedures for adding, subtracting, multiplying, and dividing fractions. (This includes connections to grades 6–8 mathematics.)
- Understanding the connection between fractions and division, $\frac{a}{b} = a \div b$, and how fractions, ratios, and rates are connected via unit rates. (This includes connections to grades 6–8 mathematics. See the Ratio and Proportion Progression for a discussion of unit rate.)" (p. 28)
- "Extending the base-ten system to decimals and viewing decimals as address systems on number lines. Explaining the rationales for decimal computation methods. (This includes connections to grades 6–8 mathematics.)" (p. 27)

Geometry and Measurement. Well-prepared beginning teachers of mathematics at the upper elementary level have strong understandings of geometry and measurement. They understand core measurement concepts such as iteration, conservation, and origin and provide frameworks for connecting linear measurement with measures of area and volume (Clements & Sarama, 2014). They can make correspondences between direct measurement of properties of shapes and algebraic approaches to determine the same measures. They can harness ideas such as line symmetry and reflection to solve problems and as tools for thinking about fractions, area, and proportions. Lehrer and Slovin (2014) noted the importance of understanding that

- "Transforming objects and the space that they occupy in various ways while noting what does and does not change provides insight into and understanding of the objects and space." (p. 8)
- Measuring attributes is one way to analyze and describe geometric shapes.
- Classifying properties of objects helps one investigate relationships between types of objects.

Well-prepared beginning teachers of mathematics at the upper elementary level can engage meaningfully in these mathematical topic areas through the mathematical practices. For example, these teachers are not only able to use measurement tools and construct figures but are also skilled in describing how to select appropriate tools. They know how to highlight the decision-making process for students, explaining how tools are used to provide accurate information. For example, well-prepared beginners in the upper elementary grades know how to use a protractor *and* also know how to talk about and show the use of this measuring tool in such a way that they could help students work through the array of challenges that often arise when using a protractor. Vignette[3] 5.1 showcases struggles that students sometimes face when learning to measure angles.

[3] This chapter includes vignettes meant to bring to life the recommendations. The vignettes serve a number of purposes, including proposing tasks that may be used with candidates for particular purposes, providing example interactions from mathematics or mathematics methods courses to exemplify effective instruction, and describing the experiences of a teacher candidate. Each vignette was chosen to highlight a particular point, but use of an isolated vignette may require surrounding context to preserve the spirit intended in the use.

VIGNETTE 5.1. MEASURING ANGLES

Ms. Hernandez is watching her students use protractors to measure a variety of angles. On a seating chart, she records what she observes small groups of students doing and key points she hears them discussing. She also sketches what the students' protractors look like in relation to the angle being measured. Some of the information she records includes the following:

Group 1, Donte: "We are doing everything right, but now there are two numbers, 20 and 160, on the protractor. Which number do we use?" (see image below)

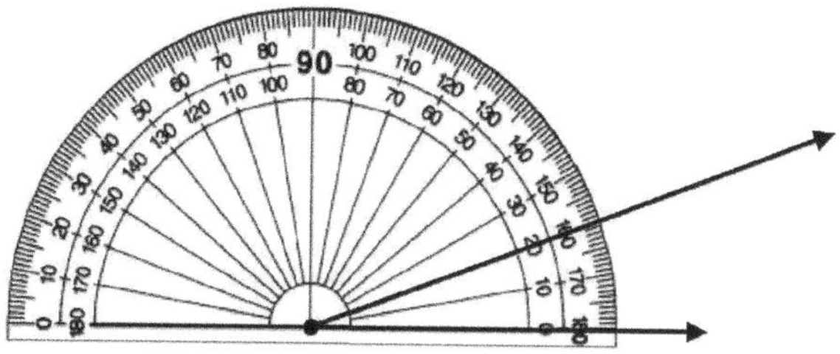

Group 2, Armando, speaking to Group 1: "Move your protractor and just look at the angle. Is it bigger or smaller than a square corner? If the angle is smaller, you pick the smaller number because a square corner is 90 degrees."

Group 3, Vince: "Where are you supposed to put the little circle?" J'La: "You put it where both of the arrows start."

Group 4, Maribel: "When I put the protractor on the bottom arrow, I get 5 inches. Is that how long the angle is?" (see image below).

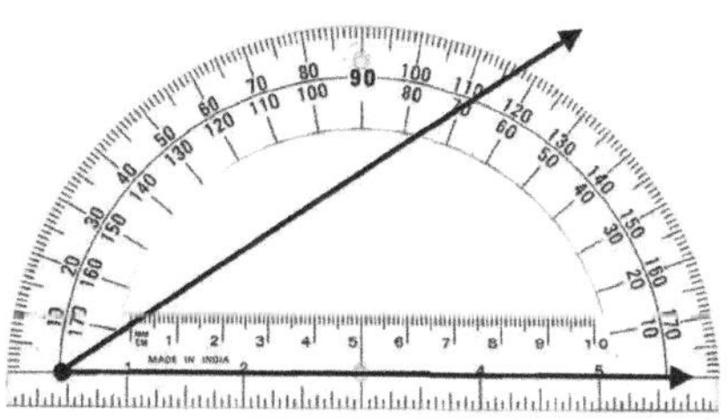

Group 5, Cristina: "This angle is too short to measure. The lines don't reach the numbers on the protractor". Diego draws the line from the angle to the scale on the protractor and says, "See, you just make the sides long enough to get to the numbers."

Ms. Hernandez anticipated many of the challenges her students encountered and wants them to collectively make sense of how to effectively use the protractor. Using her notes, she lists the students' questions and big ideas on the front board. Then, from her observations, she tentatively plans whom she will ask to share in the whole-class discussion; she will include students who present challenges for the class to think about and also students she could strategically call on to contribute ways of thinking that could move the discussion forward. She is ready to assist students in using the document camera to show how the protractor and angle looked in different situations, to support the precise use of mathematical language, and to involve many students to build on, question, and critique what is said.

Another example of well-prepared beginners' use of mathematical practices in this area is through the ways in which they are able to attend to precision. They know the affordances of referring to a square as a rectangle, rhombus, parallelogram, or quadrilateral and can generate contexts that require measurement of time to the nearest hour, minute, or tenth of a second. The well-prepared beginner also knows that decisions about precision will affect the selection and use of particular tools or representations. For example, if one is measuring to the nearest inch, she has no real need to use a ruler partitioned to show 16ths-of-an-inch increments. If weighing a very light object in a science experiment, one needs something more sensitive than a bathroom scale.

Table 5.4 lists items from the *MET II* report (CBMS, 2012) related to geometry and measurement in the upper elementary grades.

TABLE 5.4. CONNECTIONS TO *MET II* (CBMS, 2012) RELATED TO GEOMETRY AND MEASUREMENT IN THE UPPER ELEMENTARY GRADES

MET II describes the following essential ideas related to geometry and measurement:

- "Understanding geometric concepts of angle, parallel, and perpendicular, and using them in describing and defining shapes; describing and reasoning about spatial locations (including the coordinate plane).
- Classifying shapes into categories and reasoning to explain relationships among the categories.
- Reason about proportional relationships in scaling shapes up and down." (p. 30)
- "The general principles of measurement, the process of iterations, and the central role of units: that measurement requires a choice of measureable attribute, that measurement is comparison with a unit and how the size of a unit affects measurements, and the iteration, additivity, and invariance used in determining measurements.
- How the number line connects measurement with number through length...
- Understanding what area and volume are and giving rationales for area and volume formulas that can be obtained by finitely many compositions and decompositions of unit squares or unit cubes, including formulas for the areas of rectangles, triangles, and parallelograms, and volumes of rectangular prisms...
- Using data displays to ask and answer questions about data. Understanding measures used to summarize data, including the mean, median, interquartile range, and mean absolute deviation, and using these measures to compare data sets." (p. 29)

Algebraic Thinking. Well-prepared beginning teachers of mathematics at the upper elementary level have strong understandings of algebraic thinking, including the following concepts and topics (Blanton, Levi, Crites, & Dougherty, 2011):

- The fundamental properties of arithmetic, such as the commutative, associative, and distributive properties, also hold for algebra.
- Equations represent an equivalent relationship.
- Variables can be used to describe mathematical ideas and may have different meanings.
- Quantitative reasoning helps one make generalizations about relationships.

Well-prepared beginners understand students' intuitive strategies and how those strategies build on and connect with the properties of operations and other algebraic concepts (Carpenter, Franke, & Levi, 2003). They fluently use mathematical symbols and conventions to express mathematical ideas. They can readily translate and contextualize symbolic representations of phenomena as well as notice mathematical relations and patterns within real-life and problem contexts that can be expressed more generally or abstractly. They are well positioned to help students engage in mathematical practices in which thinking quantitatively is linked with opportunities to reason abstractly. Representations such as drawings, schemas, and equations are vehicles that students can use to carry their thinking into more abstract realms. While maturing, upper-elementary-grade students are increasingly able to record their thinking, so beginning teachers make the recording of thinking and reasoning an integral part of instruction and assessments at these grade levels.

Well-prepared beginners make connections between algebraic representations and graphs and tables of the same situations as well as discuss the relative advantages of different representations. They use graphs and tables to give meaning to algebraic expressions and can draw students' attention to the elegance and power of an algebraic expression to model the mathematics of a situation. They know when and how an algebraic representation can be used both to capture the logic of a numerical pattern and to show the mathematical structure of a situation.

Table 5.5 lists items from the *MET II* report (CBMS, 2012) related to algebraic thinking in the upper elementary grades.

TABLE 5.5. CONNECTIONS TO *MET II* (CBMS, 2012) RELATED TO ALGEBRAIC THINKING IN THE UPPER ELEMENTARY GRADES

MET II describes the following essential idea related to algebraic thinking:

- "Recognizing the foundations of algebra in elementary mathematics, including understanding the equal sign as meaning 'the same amount as' rather than a 'calculate the answer' symbol." (p. 26)

STANDARD C.2. PEDAGOGICAL KNOWLEDGE AND PRACTICES FOR TEACHING MATHEMATICS

Well-prepared beginning teachers of mathematics have foundations of pedagogical knowledge, effective and equitable mathematics teaching practices, and positive and productive dispositions toward teaching mathematics to support students' sense making, understanding, and reasoning.

Indicators include

C.2.1. Promote Equitable Teaching
C.2.2. Plan for Effective Instruction
C.2.3. Implement Effective Instruction
C.2.4. Analyze Teaching Practice
C.2.5. Enhance Teaching Through Collaboration With Colleagues, Families, and Community Members

Teaching mathematics entails not only knowing the mathematics but also knowing how to design and implement rich mathematics-learning experiences that advance students' mathematical knowledge and proficiencies. Effective teachers are skilled in their use of high-leverage mathematics-teaching practices and use those pedagogical practices to guide both their preparation and enactment of mathematics lessons. The development of these content-focused skills and abilities form the core of work in the preparation of mathematics teachers for the upper elementary grades. In the following section, we elaborate on the knowledge and pedagogical practices specific to the teachers for the upper elementary grades.

UE.2. Pedagogical Knowledge and Teaching Practices

> Well-prepared beginning teachers of mathematics at the upper elementary level develop pedagogical knowledge and practices to cultivate students' mathematical proficiency, including such components as conceptual understanding, procedural fluency, problem-solving ability, and facility with the mathematical processes essential for learning. [Elaboration of C.2.2 and C.2.3]

Central to any efforts to deliver high-quality instruction to each and every student is the development of pedagogical knowledge (Grossman, 1990; Shulman, 1986). Whereas Shulman pointed to two categorical groupings of pedagogical knowledge, general pedagogical knowledge and pedagogical content knowledge (PCK), both Ball et al. (2008) and Sowder (2007) made more specific connections to the role of PCK when one considers the unique components and elements essential to teaching mathematics. The well-prepared beginning teacher at the upper elementary level can engage students with the mathematics underlying the standard algorithms that are taught at these grades, including providing effective tools and ways to have students generate procedures themselves through multiple experiences. They help upper elementary students learn to select appropriate tools and use them to engage in mathematical practices such as seeing patterns and structure. Well-prepared beginners also are fully cognizant of the common errors and naive conceptions that emerge from not only these procedures with whole numbers, fractions, and decimals but also from the larger concepts in which they are embedded. The deeper the mathematical background of the well-prepared beginning teacher, the greater her or his potential for showing pedagogical sophistication (Holm & Kajander, 2012). Holm and Kajander suggested that when dividing a whole number by a fraction, the teacher should be able to choose an appropriate problem to clearly illustrate the repeated-subtraction or measurement-division approach, the strongest visual model, the most salient way of discussing the linkages to prior knowledge about division of whole numbers and a discussion of how to interpret the result.

The representations, notation, strategies, and language that are used in the classroom drive upper elementary grades students' understanding of procedures and concepts. Well-prepared beginning teachers align with conventions for proper notation, (e.g., distinguishing between a multiplication symbol and the variable *x*) and precise language (e.g., using the verb *regroup* rather than *borrow*) so that the message to students across all three grades is consistent, providing a smooth path toward building on prior knowledge in meaningful ways that last. They recognize that rather than teaching rules or shortcuts (e.g., just append a zero at the end of a number when multiplying by 10) that are taught but are applicable for only a short time (or even not very well at all), they can use more effective instructional strategies to support students in identifying patterns and identifying constraints or boundaries of the usage of those rules when they emerge. Discussing boundaries is particularly important in upper elementary grades, where students see that rules they may have learned about whole numbers do not apply to fractions and decimals (e.g., the longer the number, the larger the number) (Karp, Bush, & Dougherty, 2014). Well-prepared beginners also understand that short cuts such as searching for key words are not effective and that, instead, word problems require attention to reading-comprehension strategies. Vignette 5.2 describes a beginning teacher's work with students who were struggling with solving word problems.

VIGNETTE 5.2. STUDENTS' USE OF A KEY-WORDS STRATEGY TO SOLVE WORD PROBLEMS

Ms. Morgan was working with a small group of third graders who were having trouble solving multiplication word problems. She asked the students to meet to discuss their strategy use.

Nela was asked how she decided to use addition to solve the problem "There are three baskets of apples on the table. Each basket contains six apples. How many apples are there in all?"

Nela responded that she saw the words "in all," and that meant that the numbers listed in the problem should be added. She arrived at an answer of 9.

For the problem "Each student was given an equal share of stickers. If there are 25 stickers and 4 students, how many stickers will each student receive?" Rory said, "You use the word "each," and then you know to multiply—so they each get 100 stickers."

At this point, Ms. Morgan realized that both students were describing the use of an ineffective key-words strategy, possibly learned in previous grades. These rules or shortcuts that may have started in the primary grades were now causing serious issues. As in this case, sometimes students are mistakenly encouraged to skim through a word problem and locate the key words as a strategy to quickly choose an operation to solve the problem and then use the number(s) from the problem to carry out that operation.

Ms. Morgan had seen, in other classrooms, lists of key words that linked particular words with corresponding operations, for example, "each = multiply," and so on. But as Nela and Rory demonstrated, these words frequently do not accurately indicate the operation that corresponds with the problem. (Also, the key-word strategy cannot be used for problems that have no key words or with multistep problems). Ms. Morgan decided to show the students three word problems with the same key words but that would be successfully solved using different operations to illustrate the pitfalls and limitations of the key-words approach. She then transitioned to an annotation approach in which one student comes to the document camera and uses the suggestions of other students to mark the word problem with highlighting and written comments to identify the important information. She next moved the group to acting out the problems using the data from the annotations with paper plates and counters. The discussion then centered on the actions and how those actions relate to the meaning of the operation selected and how problems can be sorted by their structures.

Well-prepared beginning teachers focus on sense making and reasoning when they prepare students to grasp the full meaning of a problem by comprehending the entire situation and trying to use structures, such as schema, properties of the operations, and representations to come to a reasoned solution.

Also, well-prepared beginners support the learning of each and every student. This approach is particularly important in Multi Tiered Systems of Support (MTSS) such as Response to Intervention (RtI), because students are usually identified for formal special education services starting in the third grade. This process requires the careful assessment of students to pinpoint their strengths and gaps so that instruction and interventions can be targeted, whether for students with disabilities or for students who may be identified as *gifted* with a high interest in or a talent for mathematics. With reference to emerging multilingual learners, the well-prepared beginning teacher incorporates the appropriate linguistic practices and strategies needed, including home-language connections and relevant academic language and discourse practices to support students when they move to more complex mathematics vocabulary. Instruction builds on relevant contexts and the need to build on students' lived experiences in and out of the school setting. All this knowledge about, and emphasis on, teaching individual learners precludes the use of curriculum interventions via generic computer programs, basic worksheets, or Internet searches for merely attractive or fun ideas that do not support the development of significant mathematical thinking.

Although all well-prepared beginning teachers strive to align mathematical concepts across the grades, this alignment is particularly crucial for teachers of upper elementary grades who bridge work in primary grades and later work in such courses as Algebra I. A pressing challenge is teaching in ways that support the development of mathematical ideas over time while resisting the practice of teaching only the mathematics that appears in the standards for one's own grade level. For example, well-prepared beginning teachers of Grade 5 invest in knowing middle level content so that they are positioned to support students' readiness even when some of those ideas are not well represented in the fifth-grade standards. The idea of continuity of development certainly applies to teaching across upper elementary grades. For example, the responsibility for the use of number lines is represented most strongly in third-grade standards. Well-prepared beginning teachers in Grades 4 and 5 build on the development and use of the number line even though it is not specifically articulated in the standards for their grade levels. In sum, well-prepared beginners have strategic understandings of the trajectory of the representations used and take responsibility for meeting grade-level standards and reinforcing what came before (e.g., the meaning of the equal sign) and what is still to come in later grades (e.g., fifth-graders' more sophisticated use of vertical and horizontal number lines for locating points on a coordinate graph). Finally, teachers of the upper elementary grades likely have some students who need help on early childhood content and some who are ready to learn middle level content.

UE.3. Tools to Build Student Understanding

> Well-prepared beginning teachers of mathematics at the upper elementary level effectively use technology tools, physical models, and mathematical representations to build student understanding of the topics at these grade levels. [Elaboration of C.1.6 and C.2.3]

Well-prepared beginners know when to use different manipulatives and various technologies to support students in developing understanding of mathematical concepts and to create opportunities for collective consideration of mathematical ideas such as multiplication, fractions, area, volume, and coordinate geometry. They judiciously select particular representations on the basis of mathematical considerations, knowledge of their students, and other relevant factors. For example, to develop deep understandings of fractions, students must flexibly use three representations: area, linear measurement, and set models. The set model is the most complex of the three representations, so well-prepared beginners begin fractions modeling using area models and linear measurement models that connect to the number line before using the set model. Furthermore, they flexibly and resourcefully think about what representations are available in their current classrooms, schools, and wider communities; they advocate for resources to enhance their abilities to convey mathematical ideas for students to explore and discuss. This consideration of resources might include helping students' utilize calculators responsibly, giving them access to operations with large numbers and decimals that would be extraordinarily cumbersome to calculate by hand.

Well-prepared beginners also understand that meaning is not inherent in a tool or representation but that it needs to be developed through a combination of exploration, carefully orchestrated experiences, and explicit dialog focused on meaning-making (Ball, 1992). As a result they support students' developing connections between these representations, attending to links between and among equations, situations, manipulatives, tables, and graphs, using various tools including technology.

UE.4. Assessment to Promote Learning and Improve Instruction

> Well-prepared beginning teachers of mathematics at the upper elementary level learn to use both formal and informal assessment tools and strategies to gather evidence of students' mathematical thinking in ways appropriate for young learners, such as the use of observations, interviews, questioning, paper-and-pencil and computer-based tasks, and digital records, including audio and video. [Elaboration of C.3.1]

Well-prepared beginners recognize the many valued mathematical-learning outcomes that need to be assessed. They do not focus on particular outcomes to the detriment of gaining insights into others. For instance, in a unit on geometric measurement, they assess more than students' application of learned formulas; they also examine outcomes such as students' understanding of the concept of area, ability to use mathematical tools such as protractors, and attention to precision when they measure the volume of a prism. They seek to assess valued learning outcomes such as engagement in mathematical practices and mathematical dispositions, even when routes to assessing them may not be straightforward.

Well-prepared beginners utilize multiple ways to assess learning outcomes. For instance, when focusing on students' fluency with multiplication facts, they know that timed tests are not the only, first, or necessarily best approach. They recognize that fluency has several components and that timed tests do not support ability to assess strategy use, efficiency, or flexibility. They recognize that they may get a sense of students' accuracy, but primarily accuracy under the time pressure of speed. Well-prepared beginning teachers are fully aware of the negative outcomes of timed tests, which include movement away from number sense and mental computation and toward planting a seed for a negative attitude toward the study of mathematics.

Instead of relying on timed tests, the well-prepared beginner appreciates the value of looking at individual performance on assessments to pinpoint the strengths of students who are struggling (e.g., two or more grades below their peers). Using diagnostic interviews and other individualized assessments of students' thinking, they can find the gaps in foundational knowledge from previous grades as well as position instruction near the point at which students are strong in their understanding. In this way the movement forward is not in fits and leaps (as would be would be with a more gross measure of student performance in a large-scale assessment) but targeted to specific needs and built on sound footing from the learner's perspective.

STANDARD C.3. STUDENTS AS LEARNERS OF MATHEMATICS

Well-prepared beginning teachers of mathematics have foundational understandings of students' mathematical knowledge, skills, and dispositions. They also know how these understandings can contribute to effective teaching and are committed to expanding and deepening their knowledge of students as learners of mathematics.

Indicators include

C.3.1. Anticipate and Attend to Students' Thinking About Mathematics Content
C.3.2. Understand and Recognize Students' Engagement in Mathematical Practices
C.3.3. Anticipate and Attend to Students' Mathematical Dispositions

Effective teachers understand how students' mathematical ideas develop and how to apply such understandings to every aspect of teaching. There is much to learn about students' mathematical thinking, their engagement in mathematical practices, and their mathematical dispositions. In the following section, we elaborate on the knowledge and pedagogical practices needed by well-prepared beginning teachers of mathematics at the upper elementary level.

UE.5. Students' Sense Making

> Well-prepared beginning teachers of mathematics at the upper elementary level nurture students' proficiency with, and sense making of, mathematical ideas, processes, and practices. [Elaboration of C.3.1 and C.3.2]

To be well prepared for teaching mathematics in upper elementary grades, beginning teachers are ready to support students in developing increasingly sophisticated, and at times increasingly abstract, notions of mathematical ideas. Students in this grade-bands build on their additive thinking to develop the more sophisticated multiplicative thinking. Students explore the nature of fractions and decimals as well as operations involving them. They build on insights from their work with whole numbers and at times must try to avoid overgeneralizing lessons learned. They notice and describe more complex properties of shapes than they recognized previously and can express measurements of shapes in multiple ways, including with algebraic formulas. Fostering these advancements requires that beginning teachers know how mathematical ideas can progress, drawing on knowledge from research on learning progressions. They realize that using concrete, semi-concrete, and abstract representations involves overlap and integration and that their students need opportunities to revisit and try out ideas even while abstractions are developing. They help students understand the logic that makes procedures meaningful and the power and elegance mathematical conventions hold for working collectively on mathematics. Developing this understanding requires that beginning teachers actively monitor the evolution of students' ideas, be aware of likely misconceptions, and be open to understanding unique ways that students might use to express characteristics and generalizations. Beginning teachers can tailor instruction in ways that build on what students understand and consistently encourage students to stretch their mathematical thinking, such as their willingness to make conjectures or describe mathematics ideas and objects in depth.

Well-prepared beginning teachers help students become more fluent with mathematical ideas, knowing that fluency can free mental space students need to grapple with complex mathematical ideas. One commonly held goal of teachers in upper elementary grades is to support the learning of basic facts. Well-prepared beginners must be sensitive to the negative effects of practices commonly used to enhance skill with such facts, such as timed tests. Skills such as fast retrieval of multiplication facts should not be developed at the expense of sense making or developing productive dispositions toward mathematics. These teachers use methods of supporting students' need to make sense of ideas while also developing greater fluency and proficiency.

Well-prepared beginners in this grade-band nurture personal and public engagement in mathematical practices. They understand that students' mathematical identities and perceptions of mathematical status are likely to influence their participation, so they consistently work to engage all students in mathematical discussions involving explanation and critique as well as encourage the belief that each student can make valued contributions. They know that students in upper elementary grades can engage meaningfully in mathematical practices, and they build these practices on the foundations that students bring. For instance, students may believe that mathematical facts are established by the teacher, the textbook, smart classmates, or even popular consensus (i.e., by voting). Beginning teachers help students unpack the limitations of these notions while also engaging students in more robust forms of argument and proof.

Well-prepared beginners attend to the mathematical dispositions of their students. In this grade-band, students can feel greater empowerment to investigate mathematical ideas and independently address mathematical problems. Unfortunately, at this time students may, instead, shift to seeing mathematics as a collection of rules and procedures in which they lack facility and interest. Well-prepared beginners have an emerging repertoire of ways to nurture productive dispositions toward mathematics. For example, when seeing a student's expression of frustration with a particular idea or practice, a teacher might tell the student that mathematicians often struggle to solve problems or compliment the student for making several attempts to solve the problem, drawing attention to the mathematical practices and processes that are valued in mathematics.

Well-prepared beginning teachers of upper-elementary-grades students know that their students can skillfully and reflectively engage in mathematical work. As a result, they establish routines and provide students with tools to use to assess their mathematical thinking and the mathematical products they produce. This practice goes well beyond directives to "check your work." Well-prepared beginners help students express their questions and fine-tune their resources for help. They work with students to develop a shared sense of such components of high-quality work as graphs or explanations. They help students understand that accuracy and speed are not the only, and often not the best, measures of quality. They build a classroom culture in which students can provide mathematically useful feedback to their peers and the teacher and the textbook are not viewed as the only sources of mathematical validation.

STANDARD C.4. SOCIAL CONTEXTS OF MATHEMATICS TEACHING AND LEARNING

Well-prepared beginning teachers of mathematics realize that the social, historical, and institutional contexts of mathematics affect teaching and learning and know about and are committed to their critical roles as advocates for each and every student.

Indicators include

C.4.1. Provide Access and Advancement
C.4.2. Cultivate Positive Mathematical Identities
C.4.3. Draw on Students' Mathematical Strengths
C.4.4. Understand Power and Privilege in the History of Mathematics Education
C.4.5. Enact Ethical Practice for Advocacy

Effective teachers connect with students and their families. They build on students' ways of knowing and learning and attend to students' cultures, races and ethnicities, languages, genders, socioeconomic statuses, abilities, and personal interests. In the following section, we elaborate on the knowledge about social contexts of mathematics teaching and learning specific to the teachers for upper elementary grades.

UE.6. Ethical Advocates for Students

> Well-prepared beginning teachers of mathematics at the upper elementary level understand their roles as ethical advocates for elementary-grades students to have access to and advance in mathematics that cultivates positive mathematics identities and connects to students' mathematical thinking and lived experiences; these teachers build partnerships with families and communities and work to eliminate institutional and curricular barriers to learning. [Elaboration of C.4.1]

As ethical advocates for students, teachers in upper elementary grades play crucial roles in cultivating and sustaining positive learning environments that promote productive dispositions, including positive mathematics identities of students. This advocacy role includes eliciting and building on students' multiple mathematical knowledge bases (Turner et al., 2012) to expand students' thinking about new mathematical domains such as rational numbers (e.g., fractions) formally introduced in the content standards while simultaneously connecting those mathematical concepts to out-of-school experiences in families and communities that leverage these mathematical concepts. Tasks in a lesson may be focused on investigating fraction concepts used in building a community garden; analyzing public-park designs for fun, access, and safety considerations; fair sharing of resources like food and screen time; modifying a recipe for larger servings; or using tools such as a tape measure for home repairs.

Well-prepared beginning teachers in upper elementary grades understand that, with increased conceptual knowledge and procedural fluency, students have opportunities to critically analyze strengths and limitations of algorithms and representations, some of which may come from parents and grandparents schooled in different parts of the world. Well-prepared beginning teachers of upper elementary grades build on this cultural knowledge, seeking assistance from family and community members to clarify unfamiliar algorithms and possible linguistic translations when needed to build robust understandings of the algorithms and the connections to other symbolic notations and underlying concepts.

As an ethical advocate for students in upper elementary grades, well-prepared beginning teachers understand that developing positive relationships and trust with families about mathematics takes time and multiple opportunities. This means effectively communicating a positive mathematical vision for their child and creating opportunities to dialogue with parents and families about mathematics learned inside and outside the classroom. Funds of knowledge surveys and family outreach activities including home visits, faith-based center collaborations, community math workshops, and community math walks provide rich contexts to generate tasks that excite students and customize the mathematics curriculum (Aguirre et al., 2012; Civil & Bernier, 2006).

An ethical advocate for students in upper elementary grades clearly and respectfully communicates with families about their children's learning progress. In upper elementary grades, state testing and other assessment indicators are often formally used to describe students' learning progress and drive placement decisions for intervention purposes. Well-prepared beginners can provide families with holistic pictures of their children's learning progress, using combinations of student work and assessment data to help develop action plans to identify strengths and areas of growth to promote mathematical learning.

The use of standardized-test scores starting in the third grade greatly shapes instruction and the ways mathematics learning is assessed and communicated to multiple stakeholders. Often computational fluency is emphasized over conceptual understanding with the introduction of timed procedural-fluency tests. In addition, mathematical argumentation and complex problem solving are underemphasized, given the multiple-choice and short-response items often reflected in standardized tests. Unfortunately, this hyper-focus on testing has exacerbated deficit language about students with the use of terms such as *low, below basic, bubble,* and *gifted* to refer to student learning. Schools are also not immune to being labeled with general terms such as *failing.* As an ethical advocate for students in upper elementary grades, well-prepared beginning teachers are aware of this political landscape, reject the use of deficit language to describe students, and employ specific strategies in the classroom, in grade-level meetings, and during work in professional learning communities. This approach provides a comprehensive and holistic account of students' mathematical progress individually and school-wide.

PART 2. ELABORATIONS OF THE CHARACTERISTICS NEEDED BY EFFECTIVE PROGRAMS PREPARING MATHEMATICS TEACHERS FOR UPPER ELEMENTARY GRADES

In this section we recommend what preservice programs need to do to prepare their students to meet the requirements specified in the previous section. Standards specific to upper-elementary-grades mathematics teacher preparation programs are defined and described, along with additional commentary and examples related to the general standards presented in Chapter 3. Programs preparing candidates to teach a broader range of grades, such as Pre-K–5, should also meet the recommendations in this document for the Pre-K–2 preparation. The other standards are listed for the sake of completeness.

STANDARD P.1. PARTNERSHIPS

An effective mathematics teacher preparation program has significant input and participation from all appropriate stakeholders.

Indicators include

P.1.1. Engage All Partners Productively
P.1.2. Provide Institutional Support

STANDARD P.2. OPPORTUNITIES TO LEARN MATHEMATICS

An effective mathematics teacher preparation program provides candidates with opportunities to learn mathematics and statistics that are purposefully focused on essential big ideas across content and processes that foster a coherent understanding of mathematics for teaching.

Indicators include

P.2.1. Attend to Mathematics Content Relevant to Teaching
P.2.2. Build Mathematical Practices and Processes
P.2.3. Provide Sustained Quality Experiences

A high-quality preparation program for beginning teachers of mathematics for upper elementary grades provides opportunities for them to learn mathematics, focusing on both the big ideas in mathematics for those grades and the mathematical processes to make sense of mathematics in preparation to teach students.

UE.7. Mathematical Content Preparation of Teachers of Mathematics in the Upper Elementary Grades

> Effective programs preparing teachers of mathematics at the upper elementary level include coursework and other experiences focused on key mathematical ideas and skills that are pivotal in those grades. [Elaboration of P.2.1]

Because well-prepared beginning teachers must have substantial mathematical knowledge and skills as well as sound mathematical dispositions, programs must include 12 credits of coursework from a mathematics department (CBMS, 2012) and other experiences that support the development of ideas and skills that are pivotal to upper-elementary-grades teaching. For quite some time, professional organizations have called for opportunities for candidates to develop mathematical perspectives on the nature of mathematics as a discipline; the evolving nature of mathematics, especially given technological advances; and the nature of school mathematics (NCTM, 1991). Although it is important for programs to continue to implement the recommendations of the *MET II* report (CBMS, 2012), programs also must assure that candidates are not merely putting in seat-time but instead are developing deep understandings of these important concepts in ways that are usable in, and crucial for, effective teaching.

Given the role that upper elementary grades teachers play in supporting the learning of the next generation of mathematicians and mathematically astute citizens, members of a broad base of professional groups have argued that consistent, serious, and focused work on mathematics must be part of elementary school teacher preparation. Teachers preparing to teach in these grade-bands must develop mathematical knowledge that not only spans the grade levels but also provides them opportunities to understand big ideas (Charles, 2005) that unify mathematics across grade-band divides.

Also crucial is that opportunities to learn mathematics are geared to the development of mathematical knowledge that is usable in teaching (Ball et al., 2008). Courses must include opportunities for candidates to engage in activities such as unpacking multiple approaches to common mathematical tasks, examining multiple representations of a particular concept, exploring mathematical ideas through real-world contexts, explaining their solution strategies, and taking up and checking their understandings of the mathematical ideas of others.

STANDARD P.3. OPPORTUNITIES TO LEARN TO TEACH MATHEMATICS

An effective mathematics teacher preparation program provides candidates with multiple opportunities to learn to teach through mathematics-specific methods courses (or equivalent professional learning experiences) in which mathematics, practices for teaching mathematics, knowledge of students as learners, and the social contexts of mathematics teaching and learning are integrated.

Indicators include

P.3.1. Address Deep and Meaningful Mathematics Content Knowledge
P.3.2. Provide Foundations of Knowledge About Students as Mathematics Learners
P.3.3. Address the Social Contexts of Teaching and Learning
P.3.4. Incorporate Practice-Based Experiences
P.3.5. Provide Effective Mathematics Methods Instructors

A high-quality preparation program for teachers of mathematics at the upper elementary level provides opportunities for them to learn to teach mathematics by participating in well-designed, mathematics-specific methods courses.

UE.8. Mathematics Methods Coursework for Upper Elementary Teachers of Mathematics

> Effective programs for preparing teachers of mathematics at the upper elementary level include at least one mathematics methods course, or the equivalent experience of 3 semester units, focused particularly on mathematics teaching and learning in upper elementary grades. [Elaboration of P.3.1, P.3.2, P.3.3, and P.3.4]

Effective programs include coursework dedicated to the development of content-specific practices, techniques, and habits of mind or dispositions that support sound mathematics teaching. The demands of teaching and learning of mathematics in the 21st century require at least one mathematics methods course focused on upper elementary grades to fully prepare new teachers with strong foundations for success. Given the goal of educating teachers of mathematics who are well-prepared, the often-challenging nature of initial teaching assignments, and the breadth and depth of practices and dispositions needed, programs that provide certification to broad grade-bands (such as Pre-K–5 or K–8) ensure that candidates have more than one mathematics methods course. At the same time, we acknowledge that differences in the designs of teacher preparation programs will shape the ways in which this goal is achieved. As a field, we need to subject these approaches to appropriate scrutiny to ensure that program candidates are actually well prepared.

Effective methods courses include opportunities for teacher candidates to engage in mathematics, learn about students' mathematical thinking and solution strategies, identify and integrate students' lived experiences and funds of knowledge into mathematics lessons, learn various pedagogical strategies discussed in Chapter 2 and use assessment to build on students' understandings and support and extend their learning.

STANDARD P.4. OPPORTUNITIES TO LEARN IN CLINICAL SETTINGS

An effective mathematics teacher preparation program includes clinical experiences that are guided on the basis of a shared vision of high-quality mathematics instruction and have sufficient support structures and personnel to provide coherent, developmentally appropriate opportunities for candidates to teach and to learn from their own teaching and the teaching of others.

Indicators include

- P.4.1. Collaboratively Develop and Enact Clinical Experiences
- P.4.2. Sequence School-Based Experiences
- P.4.3. Provide Teaching Experiences With Diverse Learners
- P.4.4. Recruit and Support Qualified Mentor Teachers and Supervisors

A high-quality preparation program for beginning teachers of mathematics for upper elementary grades must include carefully designed and sequenced clinical placements with support structures for candidates.

UE.9. Clinical Experiences for Upper Elementary Teachers of Mathematics

> Effective programs preparing teachers of mathematics at the upper elementary level develop and systematically use a collection of clinical settings that support beginning teachers' work with diverse learners and curricula and within different institutional contexts. [Elaboration of P.4.2 and P.4.3]

The settings in which well-prepared beginning teachers at the upper elementary level learn provide supportive contexts for developing professional practices that integrate attention to students and their learning of mathematics content, including routine opportunities to see, engage in, and reflect on the mathematics teaching in real settings of practice. Mentor teachers who play roles in these upper-elementary-grades settings often are responsible for instruction across multiple subjects. Mentors of beginning mathematics teachers must routinely teach mathematics and be knowledgeable about mathematics content and pedagogy, portray productive mathematical dispositions, and open their mathematics teaching as a welcome forum for the learning of the candidates. Further, program representatives in upper-elementary-grades contexts who often are asked to provide support and feedback on the teaching of many subjects must routinely observe the mathematics teaching of the candidate as well as be knowledgeable about mathematics and portray productive mathematical dispositions.

Effective programs utilize a range of settings to provide a broad array of opportunities for candidates to learn to teach. Candidates need experiences teaching students individually, in small groups, and as a whole class; teaching diverse students, including students from many racial, gender, linguistic, and socioeconomic backgrounds; and teaching students with special needs. Effective programs include opportunities for all upper-elementary-level candidates to teach mathematics during their student-teaching experiences. When candidates are placed in elementary schools that have departmentalized staffing, arrangements must be made for candidates to have opportunities to regularly plan, teach, and reflect on mathematics teaching.

Candidates also need opportunities to develop ways of working with and learning from parents, families, and community members/leaders so that they can teach mathematics in ways that draw on students' knowledge and resources, that enhance the relevance of the mathematics taught, and forge productive partnerships that enhance education. Such opportunities could include after-school mathematics clubs, back-to-school events, school-wide mathematics nights and mathematics fairs, and home visits and should help candidates develop expertise.

STANDARD P.5. RECRUITMENT AND RETENTION OF TEACHER CANDIDATES

An effective mathematics teacher preparation program attracts, nurtures, and graduates high-quality teachers of mathematics who are representative of diverse communities.

Indicators include

P.5.1. Recruit Strong Candidates
P.5.2. Address Diverse Community Needs
P.5.3. Provide Experiences and Support Structures

Although the upper elementary grades may not appear to have the teacher shortage that middle and high schools experience, mathematics is the favorite subject for too few teachers at this level and too few feel confident in teaching the mathematics content at this level. Each and every student deserves teachers who are passionate about teaching mathematics. Given the increased rigor of mathematics in the upper grades (e.g., developing fluency with rational number concepts and operations), teacher preparation programs must recruit well-prepared beginners who have a passion for and deep understanding of mathematics.

CLOSING REMARKS

Well-prepared beginning teachers of mathematics for upper elementary grades must understand mathematics and use mathematical practices and processes, develop strong mathematical dispositions, and use mathematical tools and technology. They must learn to plan and implement effective instruction, analyze their teaching practices, and collaborate with colleagues, families, and community members. Well-prepared beginners must understand students' mathematical thinking, their use of strategies and mathematical practices, and the development of their mathematical dispositions. Candidates must understand and be committed to an advocacy role for every mathematics learner.

Effective programs preparing teachers of mathematics at the upper elementary level develop candidates' abilities to use high-leverage, effective mathematics teaching practices (NCTM, 2014a) that require deep understanding of the mathematics they are expected to teach. Programs must include carefully designed opportunities for candidates to learn effective mathematics-specific pedagogy, to learn about students as mathematics learners, and to participate in practice-based clinical experiences that are carefully designed and sequenced. High-quality programs designed on the basis of these standards will support beginning teachers' growth toward teaching excellence.

CHAPTER 6. ELABORATIONS OF THE STANDARDS FOR THE PREPARATION OF MIDDLE LEVEL TEACHERS OF MATHEMATICS

Middle level education (typically Grades 6–8) is a very important time for students when they develop intellectually, physically, and socially. The teachers of these early adolescents must understand and be able to support their mathematics and statistics learning in ways that reflect their developmental needs and interests. This chapter puts forth elaborations and examples of the standards in Chapter 2, describing the knowledge, skills, and dispositions that well-prepared beginning teachers of mathematics at the middle level need to develop, followed by elaborations and examples of the standards in Chapter 3, describing what middle level programs need to do to ensure the effective preparation of their candidates. The elaborations in this chapter focus on those standards having specific middle level considerations; therefore, although all the standards in Chapters 2 and 3 apply to middle level mathematics teacher candidates, not all require elaboration. Vignettes are included to highlight particular points and provide more detailed examples; use of an isolated vignette may require surrounding context to preserve the spirit intended for the use of the vignette. The expectations described in the middle level elaborations apply to all candidates who might teach mathematics to middle level learners. These candidates comprise those seeking mathematics certification at the middle level as well as those seeking certification that *includes* the middle level mathematics teaching, including Pre-K–8 teachers, Grades 7–12 mathematics teachers, K–12 special education teachers, and K–12 ESL specialists.

Having a coherent, well-articulated mathematics curriculum across Grades Pre-K–12 requires middle level mathematics teachers who are knowledgeable not only about the mathematics they are teaching but also about the mathematical content that is developed prior to and following the middle level years. In the middle school years, students transition from number to number systems, experience mathematical ideas more abstractly than previously, and develop foundational ideas related to algebra and geometry that are explored further in high school, college, and careers. The middle level years should not be defined as preparing students for high school but rather defined by coherent, relevant, meaningful experiences that develop competence and confidence in every middle level learner. Thus, well-prepared teachers of mathematics at the middle level reflect the skills and dispositions outlined in the *Middle Level Teacher Preparation Standards* (2012) from the Association for Middle Level Education (AMLE), the association that collaborates with NCTM, AMTE, and CAEP in middle school teacher preparation and accreditation. Table 6.1 provides a list of the middle level elaborations of the standards articulated in Chapters 2 and 3.

TABLE 6.1. ELABORATIONS OF CANDIDATE AND PROGRAM STANDARDS FOR MIDDLE LEVEL TEACHERS OF MATHEMATICS

Part 1. Candidate Knowledge, Skills, and Dispositions	
ML.1. Essential Understanding of Mathematics Content and Practices	Well-prepared beginning teachers of mathematics at the middle level have solid and flexible knowledge of relevant mathematical concepts and procedures from the middle level curriculum, including connections to material that comes before and after middle school and the mathematical processes and practices in which their students will engage. [Elaboration of C.1.1 and C.1.2]
ML.2. Content Progressions for Middle Level Learners	Well-prepared beginning teachers of mathematics at the middle level understand content progressions and the ways in which students develop mathematical content over time. [Elaboration of C.1.4]
ML.3. Strategies to Support Early Adolescents	Well-prepared beginning teachers of mathematics at the middle level use strategies to support a range of early adolescent learners and engage other educational professionals within their settings to support student learning. [Elaboration of C.2.1 and C.2.5]
ML.4. Meaningful and Interdisciplinary Contexts	Well-prepared beginning teachers of mathematics at the middle level understand how to engage middle level learners in meaningful and interdisciplinary contexts, including the use of mathematical modeling. [Elaboration of C.2.2]
ML.5. Mathematical Practices of Middle Level Learners	Well-prepared beginning teachers of mathematics at the middle level support emerging mathematical practices of middle level learners. [Elaboration of C.3.2]
ML.6. Respond to the Needs of Early Adolescents	Well-prepared beginning teachers of mathematics at the middle level understand the developmental needs of early adolescents and use this knowledge to create and implement culturally relevant mathematical experiences for their students. [Elaboration of C.4.3]
ML.7. Equitable Structures and Systems in Middle Schools	Well-prepared beginning teachers of mathematics at the middle level are aware of structures that support and inhibit opportunities for learning in schools and systems. [Elaboration of C.4.1, C.4.4, and C.4.5]
Part 2. Program Characteristics	
ML.8. Mathematics Content Preparation for Teachers of Mathematics at the Middle Level	Effective programs preparing teachers of mathematics at the middle level include content preparation aligned with *The Mathematical Education of Teachers II (MET II)* (CBMS, 2012) and *Statistical Education of Teachers (SET)* (Franklin et al., 2015). [Elaboration of P.2.1, P.2.2 and P.2.3]
ML.9. Pedagogical Preparation for Middle Level Teachers of Mathematics	Effective programs preparing teachers of mathematics at the middle level include coursework focused specifically on teaching middle level mathematics, the middle level learner, and content prior to and following middle school. [Elaboration of P.3.1, P.3.2, P.3.3, and P.3.4]
ML.10. Clinical Experiences in Middle Level Settings	Effective programs preparing teachers of mathematics at the middle level include clinical experiences in middle schools that are exemplar sites, reflecting standards for mathematics and middle level education. [Elaboration of P.4.1]

PART 1. ELABORATIONS OF THE KNOWLEDGE, SKILLS, AND DISPOSITIONS NEEDED BY WELL-PREPARED BEGINNING TEACHERS OF MATHEMATICS AT THE MIDDLE LEVEL

This section provides additional detail, commentary, and examples of the knowledge, skills, and dispositions well-prepared beginning teachers of mathematics at the middle level should have, organized by the general standards described in Chapter 2.

STANDARD C.1. MATHEMATICS CONCEPTS, PRACTICES, AND CURRICULUM

Well-prepared beginning teachers of mathematics possess robust knowledge of mathematical and statistical concepts that underlie what they encounter in teaching. They engage in appropriate mathematical and statistical practices and support their students in doing the same. They can read, analyze, and discuss curriculum, assessment, and standards documents as well as students' mathematical productions.

Indicators include

C.1.1. Know Relevant Mathematical Content
C.1.2. Demonstrate Mathematical Practices and Processes
C.1.3. Exhibit Productive Mathematical Dispositions
C.1.4. Analyze the Mathematical Content of Curriculum
C.1.5. Analyze Mathematical Thinking
C.1.6. Use Mathematical Tools and Technology

Middle level teachers need strong conceptual understandings, procedural fluency, and factual knowledge of the mathematics studied in elementary, middle, and high schools and understandings of how to develop middle level content. Middle schools often offer a range of courses, including those that offer review of elementary content and those that offer high-school-course content (e.g., algebra and geometry). Additionally, middle school mathematics has a strong focus on fluency with rational numbers and algebraic thinking, which has traditionally been taught via a narrow school mathematics curriculum in which rote procedures, rather than mathematical practices and processes, were emphasized. The content and practices related to the well-prepared beginning teacher of mathematics at the middle level are described in elaborations in this section.

ML.1. Essential Understandings of Mathematics Concepts and Practices

> Well-prepared beginning teachers of mathematics at the middle level have solid and flexible knowledge of relevant mathematical concepts and procedures from the middle level curriculum, including connections to material that comes before and after middle school and the mathematical processes and practices in which their students will engage. [Elaboration of C.1.1 and C.1.2]

In the summaries below, we describe the significant concepts that a well-prepared beginning teacher of mathematics at the middle level must know to support middle level learners. These ideas connect to and reflect *The Mathematical Education of Teachers II (MET II)* (CBMS, 2012) and *Statistical Education of Teachers (SET)* (Franklin et al., 2015) content expectations, and, therefore, we include the related essential ideas from these reports in each section.

Ratios and Proportional Reasoning. Well-prepared beginning teachers of mathematics at the middle level have the following skills and dispositions (see also Table 6.2): (a) understand a ratio as a distinct entity representing a relationship different from the quantities it compares, (b) recognize the difference between proportional and nonproportional situations, (c) experience problem situations involving more than one variable with both direct and inverse variation, and (d) know and use a variety of strategies to solve problems involving ratios and proportions (CBMS, 2012; Lamon, 2012). Well-prepared beginners understand the content that precedes ratios and proportions (multiplicative comparisons and fractions) and content that follows ratios and proportions (linear functions). They are able to select tasks that lead to the use of multiple approaches, connect to relevant contexts (e.g., financial literacy), and connect to other content within mathematics (e.g., geometry and algebra). Well-prepared beginners therefore encourage students to employ a range of reasoning strategies, including rates and scaling, ratio tables, tape or strip diagrams, double number line diagrams, and equations (proportions) (Ercole, Frantz, & Ashline, 2011; Olson, Olson, & Slovin, 2015).

TABLE 6.2. CONNECTIONS TO *MET II* (CBMS, 2012) RELATED TO RATIOS AND PROPORTIONAL REASONING AT THE MIDDLE LEVEL

MET II describes the following essential ideas for ratios and proportional reasoning:

- "Reasoning about how quantities vary together in a proportional relationship, using tables, double number lines, and tape diagrams as supports.
- Distinguishing proportional relationships from other relationships, such as additive relationships and inversely proportional relationships.
- Using unit rates to solve problems and to formulate equations for proportional relationships.
- Recognizing that unit rates make connections with prior learning by connecting ratios to fractions.
- Viewing the concept of proportional relationship as an intellectual precursor and key example of a linear relationship." (p. 41)

The Number System. Well-prepared beginning teachers of mathematics at the middle level have a unified understanding of the number system (an expected outcome for middle school students in the *CCSS-M* [NGO & CCSSO, 2010]). They understand number and the ordering of numbers in the system of rational numbers, recognize fractions, decimal fractions, and percentages as different representations of rational numbers. They understand algorithms, visual representations, and context-based problems associated with rational number operations. They are able to analyze students' algorithms and representations to provide feedback that connects conceptual and procedural knowledge. Well-prepared beginning teachers of mathematics understand the properties of the operations and know that these properties provide access to efficient or novel solution strategies. *MET II* describes essential understandings for number systems (see Table 6.3).

TABLE 6.3. CONNECTIONS TO *MET II* (CBMS, 2012) RELATED TO THE NUMBER SYSTEM AT THE MIDDLE LEVEL

MET II describes the following essential ideas for the number system:

- "Understanding and explaining methods of calculating products and quotients of fraction, by using area models, tape diagrams, and double number lines, and by reading relationships of quantities from equations.
- Using *properties of operations* (the *CCSS* term for the field axioms) to explain operations with rational numbers (including negative integers).
- Examining the concepts of greatest common factor and least common multiple.
- Using the standard U.S. division algorithm to explain why decimal expansions of fractions eventually repeat and showing how decimals that eventual repeat can be expressed as fractions.
- Explaining why irrational numbers are needed and how the number system expands from rational to real numbers." (p. 41)

Algebraic Thinking and Functions. Well-prepared beginning teachers of mathematics at the middle level have strong understandings of algebraic thinking, noticing the central role of generalization and the use of variables to represent numbers. *MET II* describes essential understandings related to expressions and equations, and for functions (see Table 6.4). In particular, these beginning teachers understand algebraic thinking as (a) the study of structures in the number system, including those arising in arithmetic; (b) the study of patterns, relations, and functions; and (c) the process of mathematical modeling (Kaput, 2008; Lloyd, Herbel-Eisenmann, & Star, 2011). Well-prepared beginners are aware that symbols such as equal signs, inequality symbols, and square-root symbols can be confusing for students and that connecting these symbols to their meanings is central to success in algebra. For example, they may ask such questions as "What answers make sense *given the context* of this problem?" or, "Does a solution set containing only the number 0 indicate having a solution?" They understand the critical importance of equivalence and approach the teaching of algebraic concepts by explicitly attending to equivalence, for example, asking students to justify methods for simplifying expressions by justifying that the new form is equivalent to the original.

Well-prepared beginners have strong conceptual and procedural understandings of linear equations, systems of linear equations, linear functions, and slope of a line. They understand that functions describe a relationship or situation in which one quantity determines another and are able to help learners understand the meaning of function and use appropriate notations and representations. Well-prepared beginners have strong understandings of how mathematical representations such as graphs, tables, and equations support and influence algebraic and functional thinking. They have strategies for integrating representations into instruction in ways that help students move flexibly among representations and connect each representation to the given context/situation.

TABLE 6.4. CONNECTIONS TO *MET II* (CBMS, 2012) RELATED TO EXPRESSIONS, EQUATIONS, AND FUNCTIONS AT THE MIDDLE LEVEL

MET II describes the following essential ideas for expressions, equations, and functions:

- "Viewing numerical and algebraic expressions as "calculation recipes," describing them in words, parsing them into their component parts, and interpreting the components in terms of a context.
- Examining lines of reasoning used to solve equations and systems of equations.
- Viewing proportional relationships and arithmetic sequences as special cases of linear relationships. Reasoning about similar triangles to develop the equation $y = mx + b$ for (nonvertical) lines." (p. 42)
- "Examining and reasoning about functional relationships represented using tables, graphs, equations, and descriptions of functions in words. In particular, examining how the way two quantities change together is reflected in a table, graph, and equation.
- Examining the patterns of change in proportional, linear, inversely proportional, quadratic, and exponential functions, and the types of real-world relationships these functions can model." (p. 43)

Geometry and Measurement. Well-prepared beginning teachers of mathematics at the middle level see the value of geometry for middle level learners; they are able to connect geometry to measurement and algebra to showcase the importance of geometry across the content while understanding that geometry itself is relevant and important for middle level learners (Sinclair, Pimm, & Skelin, 2012a). They have strategies for connecting geometry to ratios, proportions, and algebraic thinking when they explore scale drawings and transformations. Another critical connection between algebra and geometry important to middle grades mathematics is the Pythagorean Theorem; well-prepared beginners can explain why the Pythagorean Theorem is true (e.g., by decomposing a square in two ways), apply the Pythagorean Theorem, and select high-quality tasks and facilitate lessons in which their students explain and use the Pythagorean Theorem.

Well-prepared beginners understand that measurement is an important and useful mathematics content strand and therefore should be applied to authentic, culturally relevant contexts. Additionally, measurements are interrelated and must be understood and taught in ways that showcase the connections, such as connections within area formulas and connections among length, area, and volume concepts and formulas. Table 6.5 lists the essential understandings provided in *MET II* for Geometry (in which they include concepts related to measurement).

TABLE 6.5. CONNECTIONS TO *MET II* (CBMS, 2012) RELATED TO GEOMETRY AT THE MIDDLE LEVEL

MET II describes the following essential ideas for geometry (including measurement):

- "Deriving area formulas such as the formulas for areas of triangles and parallelograms, considering the different height–base cases (including the 'very oblique' case where 'the height is not directly over the base').
- Explaining why the Pythagorean Theorem is valid in multiple ways. Applying the converse of the Pythagorean Theorem.
- Informally explaining and proving theorems about angles; solving problems about angle relationships.
- Examining dilations, translations, rotations, and reflections, and combinations of these.
- Understanding congruence in terms of translations, rotations, and reflections; and similarity in terms of translations, rotations, reflections, and dilations; solving problems involving congruence and similarity in multiple ways." (p. 44)

Statistics and Probability. Well-prepared beginning teachers of mathematics at the middle level know the content described in *Statistical Education of Teachers* (*SET*) (Franklin et al., 2015) and *MET II* (see Table 6.6). They have experience with the important process of doing statistics, as explained in the American Statistical Association's (ASA's) *Guidelines for Assessment and Instruction in Statistics Education* (*GAISE*) (2007), which includes formulating questions, collecting and analyzing data, and interpreting results. Well-prepared beginners are committed and able to engage students in statistical thinking and data-analysis processes beyond computation of values, calculation of statistical measures, and rote creation of data displays that draws focus away from in-depth analysis. They have solid understandings of variability, and are able to describe the center and spread of data, understand which measures might be used in which situations, and engage students in the selection of measures to describe data. Well-prepared beginners understand, and help students understand, that different types of graphs and other data representations provide different information about the data, and, therefore, the choice of graphical representation can affect how well the data are understood. Well-prepared beginners have strong understandings of bivariate data, how to represent them, and how to help students see their connections to proportional and algebraic reasoning. Additionally, they are able to compare two data sets, noticing differences and making inferences about the two populations.

Well-prepared beginners have deep understandings of chance and the misconceptions that people have about chance (e.g., that the chance occurrence of five heads on a coin toss has no effect on whether another head will occur on the next coin toss). They appreciate that experiments, including simulations, can help students understand probability concepts as well as understand how probability is used in real-life contexts beyond games. Additionally, well-prepared beginners understand and help their students understand connections among probability, random sampling, and inference about a population.

TABLE 6.6. CONNECTIONS TO *MET II* (CBMS, 2012) AND *SET* (FRANKLIN ET AL., 2015) RELATED TO STATISTICS AND PROBABILITY AT THE MIDDLE LEVEL

MET II, *SET*, or both describe the following essential ideas for statistics and probability:

- Developing an understanding of the role of variability in statistical problem solving.
- Understanding ways to summarize, describe, and compare patterns in variability in univariate data, including frequencies, relative frequencies, and the mode (categorical data); measures of center and measures of variability (quantitative data); bar graphs (categorical data); dot plots, histograms, and box plots (quantitative data).
- Exploring patterns of association in bivariate data based on two-way tables (bivariate categorical data) and scatter plots (bivariate quantitative data).
- Understanding probability as a measure of the long-run relative frequency of an outcome, understanding basic rules of probability, and approximating probabilities through simulations.
- Understanding connections among probability, random sampling, and inference about a population.
- Comparing two data distributions and making informal inferences between two populations.

Mathematical Process and Practices. Well-prepared beginning teachers of mathematics at the middle level must be able to demonstrate mathematical practices and processes while they are learning (or relearning) the mathematics and statistics content. In these demonstrations they are able to model such things as considering various options for solving a problem; selecting an efficient strategy given the numbers or variables in the problem; using appropriate representations, models, and tools; and noticing underlying structures or patterns that can provide students insights into solving problems. Additionally, well-prepared beginners know what mathematical practices or processes are described in their state's middle level mathematics standards (e.g., the Mathematical Practices in the *CCSS-M* [NGA & CCSSO, 2010]) and know that these standards are reflected in the middle level content and are not additional, discrete topics.

Well-prepared beginners are aware of physical and technological tools that support learning of middle level mathematics. Selecting appropriate tools is a critical mathematical practice for middle level learners while they transition from more concrete content to more abstract mathematical representations. Technology tools enable teachers and students to connect differing representations of mathematical concepts and build learner knowledge within and among the representations (NCTM, 2014a). Algebraic thinking, for example, includes significant use of equations, tables, and graphs. Well-prepared beginners use virtual tools such as graphing utilities (e.g., data-graphing tools, dynamic geometry software, and electronic spreadsheets) as well as physical models and tools in various ways (e.g., for enrichment, whole-class instruction, review, new instruction) to develop students' knowledge and understandings of mathematics.

Although beginning teachers of mathematics will not see all connections between mathematical practices and content, well-prepared beginners are able to articulate strong connections between practices and content for at least one major middle school concept as well as describe some connections across the middle school topics described in Elaboration ML.1. For example, they understand the importance of contextualizing and decontextualizing with regard to mathematical problems presented through authentic situations. They apply that understanding to content instruction (e.g., of ratios) through using contextualized settings in which they have students explore the ways in which solution strategies or procedures can be derived. Vignette 6.1 provides a series of course activities as examples of support for candidates' developing deeper understanding of proportional reasoning while modeling the mathematical practices (and considering teaching practices).

VIGNETTE 6.1. A SEQUENCE OF COURSE ACTIVITIES FOCUSED ON PROPORTIONAL REASONING AND STUDENT THINKING

The following sequence of activities (based on Steinthorsdottir, n.d.) was designed to develop students' deeper understandings of proportional reasoning, number choices, and their effects in missing-value proportion problems; ways in which middle level students think about missing-value proportion problems; and strategies for posing purposeful questions to support student learning and elicit student understanding. The sequence of activities described here was implemented as a subset of activities within a 2-week unit on proportional reasoning and posing purposeful questions.

Activity 1. Solve the Potion Problem. Teacher candidates solved the Potion Problem, a missing-value proportion problem, on their own, in as many ways as they could, trying to consider ways in which middle level students might approach the problem.

> **The Potion Problem.** A potion calls for 4 drops of magical Bulbadox Juice for every two cat hairs. Neville squeezed the dropper too hard! If 44 drops of the magical Bulbadox Juice were in his pot, how many cat hairs should he use?

After working individually on ways to solve the problem, candidates shared strategies in small groups. Finally, the various strategies were shared with the whole class. For example, one strategy was written in words: There are half as many cat hairs as Bulbadox Juice, so Neville should use 22 cat hairs. Another strategy was presented numerically: $4n = 44$; $n = 11$; $2(11) = 22$. In a third explanation, candidates stated that 42 cat hairs, 2 fewer than the drops of juice, were needed. As a whole class, candidates discussed how the strategies compared, analyzing which were correct and which were alike conceptually. They came to an understanding of some generalized groupings of strategies for missing-value problems (multiplicative within measure space, multiplicative between measure space, build up (various versions), additive error, other errors). As a follow-up assignment, the candidates solved two new missing-value problems using three strategies: build up, within measure (scale factor), and between measure (invariance). Here are the two problems:

1. A group of aliens are planning a trip to Earth. For every 5 aliens they need seven Power Pouches to meet one day's food needs. If 40 aliens are traveling, how many Power Pouches will they need for each day of the trip?
2. Nina is making pillowcases to donate to a charity. She needs 18 yards of fabric for 16 pillowcases. A fabric store gave her their extra fabric. How many pillowcases can she make with 45 yards of fabric?

Activity 2. Compare Problem Numbers and Related Strategy Selection. During the next class, teacher candidates discussed The Potion Problem and the two new problems assigned with respect to numbers used in the problems and the effects of those values on potential middle level students' strategies used to solve them (i.e., the given numbers influence which strategy is selected). In the first new problem, the within-measure scale factor is a whole number whereas the between-measures space ratio is a rational number, and in the second new problem, both the within-measure-space and between-measure-space comparisons yield rational numbers. Second, candidates analyzed middle level student work from the two new problems to see to what extent the numbers in the task influenced strategy selection. Third, the candidates viewed a video clip of a researcher interviewing a middle level student solving one of the problems they had just explored and discussed the student's mathematical thinking as well as the interviewer's posing of questions.

Activity 3. Interview a Middle Level Student. Each teacher candidate conducted an interview with a middle level student, posing a set of 15 missing-value problems (with varying number choices). To prepare, they solved the problems so that they were able to select the most appropriate problems to pose to their interviewees, on the basis of what they were observing. The interview was video-recorded, and afterward, candidates analyzed the video, focusing on the students' mathematical thinking and their own posing of questions.

Several aspects of this vignette are important to note. First, this carefully sequenced series of activities, along with carefully selected tasks, provides candidates opportunities to simultaneously develop and deepen content knowledge, mathematical practices, and teaching practices. Because candidates are just beginning to anticipate student thinking, they benefit from the scaffolding of activities as in the series described here. Second, at the center of these activities are actual middle level learners. Candidates considered how middle level learners might solve problems, analyzed how they actually solved the problems, and experienced first-hand how middle level learners think about these tasks. Third, candidates enter their fieldwork prepared and with a focus on student learning. This experience is quite different from sending students to field placements in which the amount they learn and what they learn about is dependent on their assigned classroom. Because each candidate interviewed a student using the same collection of tasks, the class could collectively discuss and learn about the content, mathematical practices, and teaching practices.

ML.2. Content Progressions for Middle Level Learners

> Well-prepared beginning teachers of mathematics at the middle level understand content progressions and the ways in which students develop mathematical content over time. [Elaboration of C.1.4]

Understanding how content builds on other content is a critical component of a middle level teachers' content knowledge. As discussed in Chapter 2, available content progressions provide considerably more detail on how content develops in sophistication over time than is apparent in most standards documents. Well-prepared beginners recognize that content progressions and learning trajectories are important resources, while also recognizing that each learner is unique, with different prior knowledge and different ways he or she might approach solving problems and engaging in mathematics. For example, the progression of proportional reasoning is critical knowledge for middle level teachers. Ratios are grounded in multiplicative comparisons learned in elementary school. In middle school, students explore unitizing and rates. Students may reason about ratios and rates in many ways, from informal strategies to the use of diagrams, ratio tables, graphs, and equations. Students' ways of reasoning may make sense to them personally or reflect the way they learned about multiplicative comparisons or ratios in school previously or at home. Well-prepared beginners are able to determine and support the individual ways students reason about ratios and rates, while also helping each student deepen his or her knowledge by understanding other ways of solving ratio and proportion problems. Vignette 6.1 (above) provides an example of how course experiences can be designed to help teacher candidates deepen their understandings of student thinking related to proportional reasoning and how that reasoning grows in sophistication over time and through carefully sequenced experiences.

A solid foundation in proportional reasoning leads to developing the concept of slope as based on ratio and proportional reasoning with respect to linearity, which similarly can be conceptualized in a variety of ways. Valuing different ways of thinking about ratios, rates, proportions, slope, and so on is important in developing each student's mathematical identity.

When well-prepared middle level candidates seek to solicit students' thinking to determine where students might be on a learning trajectory, they recognize that middle level learners may feel awkward in sharing their unique thinking. Although having a collection of strategies for assessing each learner's unique understandings may be beyond the beginner's scope, well-prepared beginners understand the importance of looking for students' unique mathematical reasoning and have strategies for soliciting, understanding, and respecting the mathematical representations and explanations of their students. For example, a teacher may notice a novel strategy of a student and invite that student to project their solution, inviting the class to see how that strategy is like and different from their own and eventually asking students when they might use one strategy over another.

STANDARD C.2. PEDAGOGICAL KNOWLEDGE AND PRACTICES FOR TEACHING MATHEMATICS

Well-prepared beginning teachers of mathematics have foundations of pedagogical knowledge, effective and equitable mathematics teaching practices, and positive and productive dispositions toward teaching mathematics to support students' sense making, understanding, and reasoning.

Indicators include

C.2.1. Promote Equitable Teaching
C.2.2. Plan for Effective Instruction
C.2.3. Implement Effective Instruction
C.2.4. Analyze Teaching Practice
C.2.5. Enhance Teaching Through Collaboration With Colleagues, Families, and Community Members

Middle level learners must see the relevance and intrigue of mathematics, including its connections to the other content they are learning. Well-prepared beginning teachers of mathematics at the middle level have beginning repertoires of student-relevant contexts for each topic they teach and understand the importance of using contexts to engage students in the content. As described in the Association of Middle Level Education Standards (AMLE, 2012), beginning middle level teachers "facilitate relationships among content, ideas, interests, and experiences by developing and implementing relevant, challenging, integrative, and exploratory curriculum" (Standard 2, Element c).

ML.3. Strategies to Support Early Adolescents

> Well-prepared beginning teachers of mathematics at the middle level use strategies to support a range of early-adolescent learners and engage other educational professionals within their settings to support student learning. [Elaboration of C.2.1 and C.2.5]

Students in the middle school years are reaching a developmental stage that involves new biological and psychological experiences that may lead to sudden changes in interests and behaviors (AMLE, 2015). Well-prepared beginners are knowledgeable about the nature and developmental needs of early adolescents. For example, when children reach early adolescence, their cognitive development in mathematics sometimes far exceeds their biological or psychological development. Well-prepared beginners look for and support mathematical thinking, recognizing that a learner's behaviors might make determining what they actually know and can do challenging. Students in the middle grades, like other grades, also represent a spectrum of learners that includes students with extraordinary talents and gifts for mathematics, students with cognitive or psychological disabilities, and students for whom structures to help them meet their potentials as learners of mathematics were not in place.

Well-prepared beginners understand the specific needs of their learners, recognizing that they must hold high expectations for each and every student and employ resources such as specialists and readings to ensure that they are providing optimal environments in which each and every student can learn. They recognize mathematics-specific linguistic and cultural considerations in teaching middle level mathematics and seek ESL specialists and other resources to ensure they meet the needs of their emerging multilinguals. Well-prepared beginners seek to and are able to recognize mathematically promising students as well as create learning environments that help all learners excel. They have the disposition to seek out specialists to support and challenge students in their classrooms as well as suggest enrichment options beyond the classroom, such as clubs (e.g., The National Junior Mathematics Club [n.d.], Odyssey of the Mind [n.d.], Creative Adventures in Mathematics) and competitions (e.g., MATHCOUNTS, the American Mathematics Competition [MAA, n.d. a], Mathematical Olympiad [MAA, n.d.b]). Well-prepared beginners seek to support and challenge students with disabilities or learning challenges, accessing specialists to support their efforts. Because approximately 13% of all public school students receive special education services (National Center for Education Statistics, 2015) and because middle level mathematics learning requires significant mathematics expertise, well-prepared beginners value and seek to co-teach with special education teachers, recognizing the benefits of collaboration to support student learning. Vignette 6.2 briefly describes the roles and benefits of co-planning and co-teaching mathematics lessons.

VIGNETTE 6.2. CO-PLANNING AND CO-TEACHING TO SUPPORT EVERY STUDENT

Context. Mr. Garza is a sixth-grade mathematics teacher with 28 students including five with special needs. Mary, Angela, Morgan, and Richard have intellectual disabilities, including difficulty with reading comprehension, and Jackson is autistic and exhibits difficulty with social skills but functions well cognitively. Ms. Harris, a special education teacher, co-teaches with Mr. Garza. He has begun a unit that includes expressing one quantity, the dependent variable, in terms of the other quantity, the independent variable. This lesson is designed to help students reason about the relationship between two variables, using concrete situations and graphs. Students will receive cards with situations in words (e.g., "height of a ball thrown straight up into the air from the time it was thrown until it hits the ground") and cards with graphs to be matched with the stories.

Co-planning. The day before the lesson Mr. Garza and Ms. Harris reviewed the lesson plan and anticipated that Mary, Angela, Morgan, and Richard might struggle to read the scenarios. Jackson will likely need support in discussing why he chose his graph. Mr. W anticipates that all his students might struggle with understanding the variables that might be used for the two axes of the graphs. The teachers decide to scaffold the activity, beginning with a whole-class activity, then having students work in groups of four. Ms. Harris and Mr. Garza will be sure to monitor the learners who might struggle with reading. Mr. Garza will approach Jackson so that he can practice his explanation before the whole-class discussion.

Co-teaching. Mr. Garza begins by engaging the class in an example of a (real) ball thrown in the air. He then has students read the scenario. He asks what the variables are and how the scenario might look on a graph. He shows two graphs and asks students to tell why one matches the situation and the other one does not. During small-group time, Ms. Harris notices that many students are having difficulty with the reading, so she encourages these groups to summarize the meanings of different scenarios, giving the gist first and then adding details. All students are able to match scenarios to graphs. Mr. Garza and Ms. Harris take turns calling on a representative from each group to explain one scenario and the associated graph to the class, and each provides support feedback and comments. Jackson (prompted to rehearse his response by Ms. G) accurately and willingly shares a rationale for a match.

Well-prepared beginners know that strategies such as using multiple representations of concepts and multiple means of student action and expression are particularly important to middle level learners who are transitioning to more abstract mathematical concepts. Although they may seek guidance from instructional specialists for students identified for such services, they also are disposed to continuously find, try, and evaluate their own strategies to engage, inspire, and support every student. They recognize the critical importance of relationships for middle level learners and seek to establish relationships with each student so that they are better able to build on that student's strengths and interests to develop that student's mathematical skills and identity.

ML.4. Meaningful and Interdisciplinary Contexts

> Well-prepared beginning teachers of mathematics at the middle level understand how to engage middle level learners in meaningful and interdisciplinary contexts, including the use of mathematical modeling. [Elaboration of C.2.2]

Many contexts, interesting and accessible to middle level learners, can be investigated using mathematics. And many of these contexts can be connected to middle level content in the other disciplines (science, language arts, social studies, as well as other content). Well-prepared beginning teachers of mathematics at the middle level consider ways to design interdisciplinary instruction and are able to engage in interdisciplinary conversations, offering ideas for how important mathematics can be connected to other disciplines (AMLE, 2012). They distinguish between using mathematics as a computational tool and using mathematical reasoning or modeling, and they seek to find meaningful connections for their students.

Well-prepared beginners are knowledgeable about context-based mathematical modeling and design-based activities for middle level learners. Mathematical modeling provides not only opportunities for interdisciplinary instruction but also authentic contexts for engaging in mathematics and building mathematical understanding (Hirsch & Roth McDuffie, 2016; Usiskin, 2015). The modeling task in Table 6.7 involves a problem about the consequences of melting (de Carvalho Borba, Villareal, & da Silva Soares, 2016).

TABLE 6.7. INTERDISCIPLINARY MODELING TASK FOR THE MIDDLE GRADES

Melting of a Glacier

> The problem was to observe the percentage of reduction of a glacier, in this case, the Puncak Jaya glacier located in Indonesia . . . . This theme interested us because mankind is destroying the environment. Our hypothesis is that the glaciers have diminished to the point of almost disappearing. But to prove this hypothesis, we should consult diverse sources.

Note. Adapted from "Modeling Using Data Available on the Internet" by M. de Carvalho Borba, M. E. Villareal, and D. da Silva Soares, 2016, *Annual Perspectives in Mathematics Education: Mathematical Modeling and Modeling Mathematics,* pp. 145. Reston, VA: NCTM. Copyright 2016 by National Council of Teachers of Mathematics.

Such a task could be implemented in collaboration with teachers of English language arts (ELA), social studies, and science. The composition and production of a final report can include persuasive argument, the process of changing policies and advocating to elected officials or members of the public, and measurement and data gathering, meeting ELA, social studies, and science standards.

Well-prepared beginners recognize that a task's implementation in a classroom influences how meaningful and engaging it is for students. For example, the following list describes effective teaching practices for supporting engineering design (and mathematical modeling), a design that engages students in authentic reasoning and problem solving:

1. Pointing out limitations of the class models as a whole (e.g., if none of the models include a mechanism for motion, a teacher might request that students consider motion in a revised model);
2. Providing information students would be unable to discover on their own (e.g., explaining or contrasting the mathematical concepts of mean and median as measures of center); and
3. Encouraging individual teams of students to pursue specific design challenges to extend their models in general ways (e.g., considering how the function of the object under investigation is similar to and different from a familiar related object).

(Adapted from *Engineering in K–12 Education: Understanding the Status and Improving the Prospects* (p. 124), by the National Research Council, 2012, edited by L. Katehi, G. Pearson, and M. Feder. Washington, DC: National Academies Press. Copyright 2012 by the National Research Council.)

Middle level learners in classrooms using this engineering design were able to design functional models and data representations. Significant time is required for teachers to develop robust understandings of ways content can be integrated and ways in which engineering design and modeling activities can support mathematical content goals within any given classroom. For example, weighing the cost-benefits of paint, ensuring that lead [e.g., in paint] does not affect safety, provides a meaningful experience that addresses science and social studies. Although teachers throughout their careers will continue to learn about these relevant applications, well-prepared beginners know and have access to resources that provide engaging interdisciplinary mathematics investigations and mathematical modeling activities.

STANDARD C.3. STUDENTS AS LEARNERS OF MATHEMATICS

Well-prepared beginning teachers of mathematics have foundational understandings of students' mathematical knowledge, skills, and dispositions. They also know how these understandings can contribute to effective teaching and are committed to expanding and deepening their knowledge of students as learners of mathematics.

Indicators include

C.3.1. Anticipate and Attend to Students' Thinking About Mathematics Content

C.3.2. Understand and Recognize Students' Engagement in Mathematical Practices

C.3.3. Anticipate and Attend to Students' Mathematical Dispositions

Middle level students have their own unique understandings from their elementary schooling, life experiences, and personal preferences. The complex mathematical ideas at the middle level can almost always be approached in a variety of ways. Therefore, teachers of mathematics at the middle level must prioritize individual student reasoning as part of their planning and teaching.

ML.5. Mathematical Practices of Middle Level Learners

> Well-prepared beginning teachers of mathematics at the middle level support emerging mathematical practices of middle level learners. [Elaboration of C.3.2]

Early adolescents benefit from opportunities to explore meaningful and authentic tasks that relate to their interests and backgrounds (AMLE, 2012). Such authentic contexts provide environments from which middle level learners can further their abilities to use mathematical practices and processes. As noted earlier in this chapter, middle level mathematics is more abstract and symbolic than elementary school mathematics. Middle level learners must have regular opportunities to engage in mathematical practices and processes in a more sophisticated manner than they may have demonstrated in earlier grades. As noted by Gojak (2013), the middle grades are a time to get "messy" with mathematics. Therefore, well-prepared beginners must be able to engage their students in representing and explaining their mathematical thinking and making mathematical arguments. For example, they must be able to understand the various approaches students might use to solve problems and create environments in which strategies are discussed, critiqued, and compared.

Well-prepared beginners understand that particular teaching moves support (or inhibit) student development of mathematical practices and processes. For example, as discussed in Activities 2 and 3 in Vignette 6.1, posing questions can help to develop students' abilities to analyze problem situations, select appropriate strategies, and reason quantitatively. Well-prepared beginners also recognize that early adolescents may be self-conscious and therefore have a variety of ways for students to demonstrate their unique thinking. For example, a well-prepared beginner might collect work and project particular solutions anonymously to highlight reasoning or particular representations. Well-prepared beginners reflect on ways their own actions affect the ongoing development of student thinking.

STANDARD C.4. SOCIAL CONTEXTS OF MATHEMATICS TEACHING AND LEARNING

Well-prepared beginning teachers of mathematics realize that the social, historical, and institutional contexts of mathematics affect teaching and learning and know about and are committed to their critical roles as advocates for each and every student.

Indicators include

C.4.1. Provide Access and Advancement
C.4.2. Cultivate Positive Mathematical Identities
C.4.3. Draw on Students' Mathematical Strengths
C.4.4. Understand Power and Privilege in the History of Mathematics Education
C.4.5. Enact Ethical Practice for Advocacy

As suggested by Dewey (1910), improving the public's knowledge of mathematics and science and cultivating critical reasoning skills are important goals for education and essential for a democratic citizenry. Middle grades become important in a child's development for bringing numerical and other mathematical reasoning tools to bear on the decisions citizens must make in a democracy, even when research shows that strongly held beliefs and cultural affinities may disable individuals' use of reason and sense making to inform decisions (Kahan, Peters, Dawson, & Slovic, 2013). Mathematics, in particular, provides young, future citizens with useful tools and practical lenses to examine social and personal issues that arise throughout their lives. Middle school mathematics teachers must communicate and model the power and utility of mathematics for decision making and personal growth.

ML.6. Respond to the Needs of Early Adolescents

> Well-prepared beginning teachers of mathematics at the middle level understand the developmental needs of early adolescents and use this knowledge to create and implement culturally relevant mathematical experiences for their students. [Elaboration of C.4.3]

Well-prepared beginning teachers of mathematics at the middle level are able to cultivate the development of students' positive mathematical identities and draw on students' cultural and linguistic strengths as well as their individual interests and passions as part of the middle school mathematics instructional program. As an example, using personal-information surveys can help early-adolescent learners to examine data and measurement units that are plausible (Lovett & Lee, 2016). Similarly, proportional reasoning skills can be developed by examining socially relevant questions about equity and fairness in the contexts of consumer issues, population growth, and crime rates (Simic-Muller, 2015). Drawing on how different cultures employ mathematical ideas provides students opportunities to learn about and honor their own and other cultures as exemplified in how students from immigrant families might participate in a *culture-laden* lesson that develops "a compassionate understanding of their classmates from different backgrounds and [fosters] an atmosphere of respect, solidarity, and collaboration" (Taylor, Rehm, & Catepillán, 2015, p. 106). Additionally, students' language can be used as a resource, as illustrated in Vignette 6.3.

VIGNETTE 6.3. ATTENDING TO ALGEBRA CONTENT AND LANGUAGE IN A CLASSROOM WITH EMERGENT MULTILINGUAL STUDENTS

Rosa, a student teacher, was thrilled to be teaching with Marie, a 6th-grade teacher in a school designed specifically for recent immigrants. Students in her class represented 13 countries and 11 languages. Although numerous students were Spanish speakers, some students were the only ones speaking their native languages. Together, Rosa and Marie designed a 2-week unit on algebraic thinking, with a focus on generalizing patterns, writing algebraic expressions and equations, and connecting representations (situations, generalized rules, equations, and tables).

With the students' diverse backgrounds and need for visual supports, Rosa and Marie decided to infuse children's literature as a tool for building common background. The first book, *Two of Everything* (Hong, 1993), was a Chinese folktale that delighted the sole Chinese student and was well received by everyone. Knowing which students had stronger understandings of English and felt comfortable in front of the class, Marie asked four students to come to the front and act out the story while she read the book (she had brought some basic props from home). In the story, a magic pot is found that doubles everything that goes in it.

After reading the story, Rosa asked students to tell their shoulder partners what happened in the story. Along with other aspects of the story, they talked about the magic pot doubling whatever was dropped into it. Because the students spoke different languages, they often spoke to each other in everyday English. During planning, Marie had encouraged Rosa to build meaning for the mathematical language, using their everyday languages as a resource. During the lesson, Rosa recorded a vocabulary table on the board:

	Everyday Meaning	Mathematics Meaning
Rule		
Expression		
Sentence/Equation		

Rosa asked students to talk to their shoulder partners about the meanings of the words when they use them outside of math class. Rosa recorded their examples of the everyday meanings. For each one, Rosa referred to the pattern in the magic pot and added in the language or symbols connected to the story.

	Everyday Meaning	Mathematics meaning
Rule (words)	"Quiet" "No cell phones"	"Multiply by 2"
Expression (symbols/numbers)	"How beautiful!"	$2 \times a$ $2 \bullet a$ $a \times 2$ $2a$
Sentence/Equation (*equation* is a number sentence)	"How beautiful <u>are</u> your eyes!"	$n = 2 \bullet a$ $2 \times a = n$

After completing the table, Rosa had students say each word together. Then she explained the magic pot had changed its rules, and it would be their job to determine the new rules and to write expressions and equations for each new rule. She distributed an activity page that had several input-output tables (labeled as In-the-Pot and Out-of-the-Pot), with places to record rules, expressions, and equations. These tables were then solved and discussed.

In reflecting, both Rosa and Marie noted every student used the three identified words when they discussed the new magic-pot rules. Marie noticed the Chinese student had spoken in English without first using his translator for the first time. While the unit progressed, students continued to use appropriate language and to explore increasingly complex linear relationships across a range of situations, including generating their own stories.

This vignette provides an example of maintaining high expectations and providing strong support for students. It shows the value of co-planning and co-reflecting, and the effect this can have on the teacher candidate as well as on the students.

In the middle school mathematics classroom, early adolescent learners can examine complicated challenges they will confront at school and in their lives, using the analytical and logical tools provided by mathematics. Well-prepared beginners must be careful to avoid advocacy of a particular point of view but can examine how mathematics informs opinions and decisions about topics that are current, relevant, or of particular interest to learners (such as climate and the environment, health and human sexuality, bullying, or lotteries) and thereby empower students to use mathematics for critical thinking and authentic purposes. Well-prepared beginners realize that such careful selection and enactment of tasks shapes students' emerging mathematical identities and influences the decisions they will make in terms of continuing in mathematics, pursuing careers, and selecting college majors.

ML.7. Equitable Structures and Systems in Middle Schools

> Well-prepared beginning teachers of mathematics at the middle level are aware of structures that support and inhibit opportunities for learning in schools and systems. [Elaboration of C.4.1, C.4.4, and C.4.5]

Well-prepared beginners recognize that current systems and structures do not provide equitable opportunities for early adolescents to learn mathematics. Tracking is one such structure that is typically initiated at the middle level (Loveless, 2016). *Tracking,* the practice of grouping students in classes on the basis of perceived ability levels, leads to enrollment of some students in algebra courses in the middle grades and other students being kept from enrolling in such courses. Tracking has been demonstrated to create and reinforce social inequities because African American, Latinx, and children living in poverty are underrepresented in the accelerated tracks (Boaler, 2011; Larnell, 2016).

Different from tracking, but related, is differentiation through enrichment and acceleration. Enrichment provides opportunities within courses to deepen student understanding; acceleration is a faster pace through a curriculum. Rushing elementary learners to algebra and high school learners to calculus is generally identified as *acceleration* and is counter to ample research (Bressoud, Mesa, & Rasmussen, 2015). This practice has contributed to practices at the middle level that are inconsistent with the principles of middle level education articulated by the Association for Middle Level Education (AMLE, 2010) and runs counter to principles of equity. Both enrichment and acceleration are used in middle level to differentiate instruction. Well-prepared beginners recognize the distinctions between enrichment and advancement and recognize also that although the intent might be to provide support and challenge to all students, many efforts to provide enrichment or acceleration have resulted in inequities and denied students access to important mathematics. Well-prepared beginners realize that they can advocate for more equitable practices related to advancement. For example, they might advocate that eighth-grade Algebra I placement assessments to be based on not only high-stakes-test scores but on multiple measures based on student potential, creativity, and interests. Well-prepared beginners are committed to implementing enriched curricula, recognizing the importance of developing the mathematical practices and positive mathematical dispositions for each and every student.

PART 2. ELABORATIONS OF THE CHARACTERISTICS NEEDED BY EFFECTIVE PROGRAMS PREPARING MIDDLE LEVEL TEACHERS

This section provides additional details, commentary, and examples for what is needed in programs to effectively prepare their students to teach middle level mathematics, organized by the general standards described in Chapter 3. As noted in the introduction, these elaborations build on the expectations in Chapter 3, and focus on additional considerations specific to the preparation of teachers of mathematics at the middle level.

STANDARD P.1. PARTNERSHIPS

An effective mathematics teacher preparation program has significant input and participation from all appropriate stakeholders.

Indicators include

P.1.1. Engage All Partners Productively
P.1.2. Provide Institutional Support

Partnerships are particularly important in ensuring the effective preparation of middle level mathematics teachers. Partnering with middle schools with teachers who model middle level practices provides opportunities for candidates to experience the benefits of teaming and interdisciplinary instruction, among other things. Partnerships between colleges of education and departments of mathematics, for example, provide candidates opportunities to coordinate the content, experiences, and sequencing of middle level mathematics content, mathematics methods, and other middle level courses. Chapter 3 provides significant guidance on building effective partnerships that can support the preparation of well-prepared beginning teachers of mathematics at the middle level.

STANDARD P.2. OPPORTUNITIES TO LEARN MATHEMATICS

An effective mathematics teacher preparation program provides candidates with opportunities to learn mathematics and statistics that are purposefully focused on essential big ideas across content and processes that foster a coherent understanding of mathematics for teaching.

Indicators include

P.2.1. Attend to Mathematics Content Relevant to Teaching
P.2.2. Build Mathematical Practices and Processes
P.2.3. Provide Sustained Quality Experiences

Effective programs include content courses that address these content needs of middle level teachers of mathematics. As described in Part 1 of this chapter, middle level candidates must be prepared to teach a broad range of mathematics topics, understanding the content, progressions, relationships among content, not just at the middle level, but before and after middle level. Further, middle level candidates need to understand historical and cultural aspects of mathematics. Regardless of licensure options (e.g., K–8, 5–8, 6–12, or any other) effective preparation of middle level mathematics teachers includes opportunities to learn deeply the mathematics content that is taught at the middle level as well as mathematics content that comes before and after.

ML.8. Mathematics Content Preparation for Teachers of Mathematics at the Middle Level

> Effective programs for teachers of mathematics at the middle level include content preparation aligned with *The Mathematical Education of Teachers II (MET II)* (CBMS, 2012) and *Statistical Education of Teachers (SET)* (Franklin et al., 2015). [Elaboration of P.2.1, P.2.2 and P.2.3]

Consistent with the recommendations of *MET II* and *SET,* the following coursework is needed to prepare middle level teachers:

- At least 15 semester hours (or equivalent) of mathematics and statistics courses designed specifically for future middle level teachers, including courses that engage middle level candidates in opportunities to demonstrate the mathematical practices.
- At least 9 semester hours (or equivalent) of mathematics and statistics courses beyond the precalculus level, including at least one statistics course.

The content within these courses must address the content needs of middle level teachers of mathematics. Across the more than 24 hours of coursework, a high-quality middle level mathematics preparation program provides coursework that addresses the following mathematical concepts (content lists and course recommendations are from *MET II* [CBMS, 2012] and *SET* [Franklin et al., 2015]):

Number and Operations. Number and operations in base ten, fractions, addition, subtraction, multiplication, and division with whole numbers, decimals, fractions, and negative numbers. Possible additional topics are irrational numbers or arithmetic in bases other than ten. (6 semester-hours recommended)

Geometry and Measurement. Perimeter, area, surface area, volume, and angle; geometric shapes, transformations, dilations, symmetry, congruence, similarity; and the Pythagorean Theorem and its converse. (3 semester hours recommended)

Algebra and Number Theory. Expressions and equations, ratio and proportional relationships (and inversely proportional relationships), arithmetic and geometric sequences, functions (linear, quadratic, and exponential), factors and multiples (including greatest common factor and least common multiple), prime numbers and the Fundamental Theorem of Arithmetic, divisibility tests, rational versus irrational numbers. Additional possible topics for teachers who have already studied the above topics in depth and from a teacher's perspective are polynomial algebra, the division algorithm and the Euclidean algorithm, and modular arithmetic. (3 semester hours recommended)

Statistics and Probability. Describing and comparing data distributions for both categorical and numerical data, exploring bivariate relationships, exploring elementary probability, and using random sampling as a basis for informal inference. An effective program for middle level teachers requires not only an introductory course to statistics but also a course that includes data collection and analysis. Such an experience emphasizes active learning with appropriate hands-on devices and technology with teachers probing deeply into the topics taught in the middle grades, all built around seeing statistics as a four-step investigative process involving question development, data production, data analysis, and contextual conclusions (ASA/NCTM, 2015; CBMS, 2012; Franklin et al., 2015). (6 semester hours recommended)

Implicit in these lists is the importance of candidates' understanding the mathematical content of elementary as well as high school levels. For example, one of the two course recommendations in the area of number may be a mathematics course for elementary school teachers focused on rational numbers. Additionally, designers of a well-designed middle level mathematics program strategically consider the sequence in which courses occur, including the extent to which courses are taken prior to or concurrently with education courses, mathematics methods courses, and clinical experiences. Although having mathematical knowledge is a prerequisite to teaching mathematics, having mathematics course opportunities later in a program can have more connection to a middle level candidates' classroom teaching. A well-designed program provides at least one content-course experience that is concurrent with a candidate's clinical experience so that connections can be made between the content being learned and content that is being taught in the clinical setting.

STANDARD P.3. OPPORTUNITIES TO LEARN TO TEACH MATHEMATICS

An effective mathematics teacher preparation program provides candidates with multiple opportunities to learn to teach through mathematics-specific methods courses (or equivalent professional learning experiences) in which mathematics, practices for teaching mathematics, knowledge of students as learners, and the social contexts of mathematics teaching and learning are integrated.

Indicators include

P.3.1. Address Deep and Meaningful Mathematics Content Knowledge
P.3.2. Provide Foundations of Knowledge About Students as Mathematics Learners
P.3.3. Address the Social Contexts of Teaching and Learning
P.3.4. Incorporate Practice-Based Experiences
P.3.5. Provide Effective Mathematics Methods Instructors

As described in Chapter 3, mathematics methods courses are critical to the preparation of well-prepared beginners. For middle level teacher candidates to be well-prepared, they need middle-level-focused methods courses in which they have opportunities to apply their developing teaching skills and their knowledge of mathematics to the teaching of specific mathematical topics at the middle level. Such preparation is necessary but not sufficient for a well-prepared middle level candidate. A high-quality program also requires coursework that prepares middle level teachers of mathematics to understand the learner, the content they have learned prior to middle school, and the content they will learn after middle school.

ML.9 Pedagogical Preparation for Middle Level Teachers of Mathematics

> Effective programs preparing teachers of mathematics at the middle level include coursework focused specifically on teaching middle level mathematics, the middle level learner, and content prior to and following middle school. [Elaboration of P.3.1, P.3.2, P.3.3, and P.3.4]

For middle level teacher candidates to be well prepared, they need middle-level-focused methods courses in which they have opportunities to apply their developing teaching skills and their knowledge of mathematics to the teaching of specific mathematical topics at the middle level. In addition to this mathematics-specific, middle level experience, those preparing to teach middle level mathematics should have at least one additional intense learning opportunity focused on the middle level learner. This experience may be a second content methods course (e.g., a secondary mathematics methods course), a general middle level course, or a middle school methods course in a different discipline (e.g., middle school science methods). Having more than one course focused on middle level learners or on a second content area provide candidates opportunities to explore interdisciplinary connections as well as other aspects of the middle school and middle level learners, integrative and experiential curriculum, instructional strategies appropriate to the early adolescent, and the ways in which middle schools function.

As the name of the grade-band indicates, middle level educators teach at a transitional point in a learner's K–12 mathematics education. Additionally, mathematics experiences and learning across these years transition from concrete and visual content toward more abstract and complex content. Students also enter middle level grades with gaps in their understandings of elementary-level mathematics. As described in middle-level-candidate expectations earlier in this chapter, middle level teachers of mathematics must know what content is taught prior to and after middle school as well as know how to assess the extent that their students have learned what was previously taught and then address any learning gaps while also teaching the appropriate middle level content. Acquiring such knowledge and skills requires significant coursework and experiences with elementary and high school mathematics and mathematics teaching. Rational-number learning, for example, begins with students' elementary school introductions to this number concept, a necessary foundation to accomplish the highly developed understandings of rational number expected in middle school. In a similar way, middle level teachers of mathematics must understand the concepts in algebra, geometry, statistics, and functions that are increasingly abstract and complex and are foundational to the study of high school mathematics. This understanding may be acquired in a program through the inclusion of elementary or secondary content-for-teacher courses, additional methods courses, or attention to elementary and secondary mathematics within a series of middle level mathematics courses.

STANDARD P.4. OPPORTUNITIES TO LEARN IN CLINICAL SETTINGS

An effective mathematics teacher preparation program includes clinical experiences that are guided on the basis of a shared vision of high-quality mathematics instruction and have sufficient support structures and personnel to provide coherent, developmentally appropriate opportunities for candidates to teach and to learn from their own teaching and the teaching of others.

Indicators include

P.4.1. Collaboratively Develop and Enact Clinical Experiences
P.4.2. Sequence School-Based Experiences
P.4.3. Provide Teaching Experiences With Diverse Learners
P.4.4. Recruit and Support Qualified Mentor Teachers and Supervisors

As stated in the AMLE Middle Level Teacher Preparation and Certification/Licensure guide (2015), effective middle level teacher preparation programs place high priorities on providing and requiring early and continuing middle level clinical experiences for prospective middle level teachers. The priority given these experiences reflects the views of practicing teachers about the essential components of professional preparation programs (Wilson, Floden, & Ferrini-Mundy, 2001).

ML.10. Clinical Experiences in Middle Level Settings

> Effective programs preparing teachers of mathematics at the middle level include clinical experiences in middle schools that are exemplar sites, reflecting standards for mathematics and middle level education. [Elaboration of P.4.1]

Programs for the preparation of middle level teachers of mathematics seek placements in model middle schools. Consistent with the standards of the AMLE and expressed in the organization's landmark position paper *This We Believe* (2010), effective middle schools address the needs of middle level learners, including emphases on students helping one another to be successful and developing the abilities to contribute positively to their communities and the world. Schools at which well-prepared beginning teachers serve as interns and develop their practices through clinical experiences must exemplify practices that support the needs of middle level learners. For example, young adolescents tend to be highly curious and display broad arrays of interests (which can quickly change), are eager to learn about things they find interesting, and prefer active learning (Kellough & Kellough, 2008).

Those responsible for mathematics planning, teaching, and assessing practices within the school and the assigned classroom(s) must clearly value, advocate for, and understand the unique characteristics of young adolescents. Model middle schools also have organizational features such as interdisciplinary teams, learning environments, and time structures that contribute to learning and achievement; they avoid practices such as tracking that lower expectations for many learners. Furthermore, the school environment must be safe, inclusive, and supportive of the learners' needs. The involvement of families, businesses, and other members of the community must be evident and active. When such a middle school is not available for candidate placement, middle level candidates need opportunities to consider how such environments can be created in their own classrooms. Engaging in experiences such as viewing videos, group discussions, and personal reflections can supplement candidates' clinical placements. Additionally, if opportunities for clinical experiences in exemplar middle schools are limited, programs can create partnerships with schools to develop a cadre of middle level teachers who can help provide a reinforcing environment for what teacher candidates are learning.

To ensure that the curriculum principles unique to middle level education are addressed, programs must ensure that clinical internship or practicum placements occur not only within a mathematics classroom but also with an instructional team that involves teachers with licensure or certification in other disciplines. Having full certification in two content areas is not necessary, but well-prepared beginning teachers of middle school mathematics need to have practical experiences teaching content that complements mathematics and is part of the diverse middle level school curriculum. In particular, the integration of information-literacy skills and appropriate state-of-the–art technologies into teaching mathematics to meet the needs of all young adolescents (i.e., regardless of race, ethnicity, culture, age, appearance, ability, sexual orientation, socioeconomic status, family composition) is essential.

STANDARD P.5. RECRUITMENT AND RETENTION OF TEACHER CANDIDATES

An effective mathematics teacher preparation program attracts, nurtures, and graduates high-quality teachers of mathematics who are representative of diverse communities.

Indicators include

P.5.1. Recruit Strong Candidates
P.5.2. Address Diverse Community Needs
P.5.3. Provide Experiences and Support Structures

The recruitment and retention of high-quality teachers of mathematics for the middle grades offer unique challenges to educator preparation programs. Middle school mathematics certification may be relatively new or not required in some states. Majors or programs specific to middle level preparation are not well known, particularly among entering college students who typically sort themselves between majors or career goals tied to elementary or secondary school teaching. Often, secondary school is synonymous with Grades 9–12 or high school, and middle school focus and opportunities are missed. Related to retention, time and attention to the ability of middle school teachers to work effectively with mentors as well as with other educational specialists (e.g. counselors, social workers, special education professionals) are important both to initial preparation of teachers and during their induction periods.

As described in Chapter 3, the Mathematics Teacher Education Partnership Research Action Cluster (Ranta & Dickey, 2015) on recruitment has identified effective recruitment strategies for mathematics teachers. Here we revisit these strategies as they apply to attracting high school and college students to middle level majors, where they exist, or to programs leading to middle grades mathematics certification or licensure include.

- Offering field experiences in middle school mathematics settings with exemplary teachers
- Providing scholarships specific to middle level programs
- Promoting the need for middle grades mathematics teachers that exceeds the need for elementary teachers as well as for middle level English/language art or social studies teachers
- Highlighting the integrated and active-learning curriculum intended for middle grades learners
- Building a connection to the unique emotional and cognitive needs of middle grade learners
- Providing career counseling to elementary and secondary education students as well as to mathematics majors about major changes and certification options specific to middle school teaching

CLOSING REMARKS

Effective teacher preparation programs at the middle level must develop candidates' abilities to use high-leverage, effective mathematics teaching practices (NCTM, 2014a) that reflect the needs of early adolescents. A priority in the preparation of middle level candidates is helping them commit to such teaching practices, even if that is not how they experienced middle level mathematics. This commitment will come to fruition only when programs provide robust experiences through courses and fieldwork that illustrate the benefits that effective mathematics instruction can have on each and every middle level learner.

CHAPTER 7. ELABORATIONS OF THE STANDARDS FOR THE PREPARATION OF HIGH SCHOOL TEACHERS OF MATHEMATICS

High school mathematics teachers must have strong content knowledge, knowledge of mathematics-specific pedagogy, and much more—including knowledge about their individual students and their cultural contexts, school policies, and how to collaborate with other teachers. Only with this knowledge, will mathematics teachers be able to meaningfully support the learning of each and every student.

In this chapter, particular attention is given to ensuring that each and every high school student has opportunities to learn meaningful mathematics well in equitable and empowering learning environments. In Part 1 of this chapter, we put forth elaborations and examples of the standards in Chapter 2, describing the knowledge, skills, and dispositions that well-prepared beginning high school mathematics teachers need to develop, highlighting actions they need to take. Part 2 provides elaborations and examples of the standards in Chapter 3, describing what is needed in high school level preservice programs to ensure the effective preparation of their candidates. The chapter concludes with example approaches that programs might take in achieving the standards. A summary of the high school elaborations is given in Table 7.1. The elaborations in this chapter focus on those standards for which there are specific high-school-level considerations; therefore, although all the standards in Chapters 2 and 3 apply to high-school-level mathematics teacher candidates, not all require elaboration.

TABLE 7.1. ELABORATIONS OF CANDIDATE AND PROGRAM STANDARDS FOR HIGH SCHOOL TEACHERS OF MATHEMATICS

Part 1. Candidate Knowledge, Skills, and Dispositions	
HS.1. Essential Understandings of Mathematics Concepts and Practices in High School Mathematics	Well-prepared beginning teachers of mathematics at the high school level have solid and flexible knowledge of relevant mathematical concepts and procedures from the high school curriculum, including connections to material that comes before and after high school mathematics and the mathematical processes and practices in which their students will engage. Relevant mathematical concepts include algebra as generalized arithmetic, functions in mathematics, diagrams and definitions in geometry, and statistical models and statistical inference. [Elaboration of C.1.1 and C.1.2]
HS.2. Use of Tools and Technology to Teach High School Mathematics	Well-prepared beginning teachers of mathematics at the high school level are proficient with tools and technology designed to support mathematical reasoning and sense making, both in doing mathematics themselves and in supporting student learning of mathematics. In particular, they develop expertise with spreadsheets, computer algebra systems, dynamic geometry software, statistical simulation and analysis software, and other mathematical action technologies as well as other tools, such as physical manipulatives. [Elaboration of C.1.6]

HS.3. Supporting Each and Every Student's Opportunity to Learn Mathematics	Well-prepared beginning teachers of mathematics at the high school level understand the importance of providing each and every high school student with opportunities to learn mathematics that will enable him or her to think analytically and creatively in preparation for the workforce, college, citizenship, and life. [Elaboration of C.2.1]
HS.4. Cultivating Positive Mathematical Identities in Each and Every Student	Well-prepared beginning teachers of mathematics at the high school level draw on students' strengths to cultivate positive mathematical identities. [Elaboration of C.4.2 and C.4.3]

Part 2. Program Characteristics

HS.5. Effective Programs to Support Preparation of Teachers of Mathematics at the High School Level	Effective programs preparing teachers of mathematics at the high school level are specifically focused on preparing high school teachers of mathematics.
HS.6. Partnerships to Support Preparation of Teachers of Mathematics at the High School Level	Effective programs preparing teachers of mathematics at the high school level engage mathematicians and statisticians, school partners, and other stakeholders in supporting the growth of their candidates to become well-prepared beginning teachers. [Elaboration of P.1.1]
HS.7. Mathematical Content Preparation of Teachers of Mathematics at the High School Level	Effective programs preparing teachers of mathematics at the high school level are focused on the relevant content knowledge needed for teaching high school mathematics, including connections to material that comes before and after high school mathematics. Coursework consists of the equivalent of an undergraduate major in mathematics (including statistics) with at least three content courses particularly relevant to teaching high school mathematics and incorporating sufficient attention to a data-driven, simulation-based modeling approach to statistics. [Elaboration of P.2]
HS.8. Ethics and Values for Teaching Mathematics at the High School Level	Effective programs preparing teachers of mathematics at the high school level include multiple opportunities for candidates to develop political clarity on the profession and their advocacy roles in teaching. [Elaboration of P.3.3]
HS.9. Mathematics Methods Experiences for Teachers of Mathematics at the High School Level	Effective programs preparing teachers of mathematics at the high school level provide multiple opportunities for candidates to learn to teach mathematics effectively through the equivalent of three mathematics-specific methods courses. [Elaboration of P.3.4]
HS.10. Clinical Experiences for Teachers of Mathematics at the High School Level	Effective programs preparing teachers of mathematics at the high school level provide clinical experiences in which candidates develop teaching practices that support the learning of conceptual knowledge and mathematical practices and processes for each and every student. [Elaboration of P.3.5]

Programs preparing candidates to teach a broader range of grades than Grades 9–12 (e.g., 6–12 or 7–12) must also attend to the recommendations in this document for middle-level preparation. Preparation to teach high school does not automatically prepare a candidate to teach middle school. Such candidates need to have content courses focused on middle level mathematics content, methods courses particularly focused on teaching middle school mathematics, understanding middle school students, and significant clinical experiences at middle school. See Chapter 6 for details as well as additional commentary about preparing candidates to teach a broader range of grades embedded throughout this chapter.

This chapter includes a number of vignettes meant to bring to life the recommendations put forward. The vignettes serve a number of purposes, including proposing tasks that may be used with candidates for particular purposes, providing example interactions from mathematics or mathematics methods courses to exemplify effective instruction, and describing the experiences of teacher candidates. Each vignette was chosen to highlight a particular point, but use of an isolated vignette may require surrounding context to preserve the spirit intended by the use of the vignette.

The vignettes show the importance of well-prepared beginning teachers' possessing strong pedagogical content knowledge and knowledge of their students. Even though some of the vignettes appear to focus more on issues of access and empowerment, mathematical content is a central component of what is discussed. Teacher candidates need opportunities to think about scenarios in which each and every student is involved in reasoning and sense making of mathematics and those in which some students are not. Providing high school mathematics teacher candidates with such opportunities helps them to examine their own beliefs and to think about how their beliefs influence what instructional practices they value and use to support or unintentionally marginalize particular students. For secondary-level mathematics teacher preparation to be focused not only on mathematics content, which tends to be in the forefront, but also on affective factors that may affect students' mathematics engagement and achievement is imperative.

PART 1. ELABORATIONS OF THE KNOWLEDGE, SKILLS, AND DISPOSITIONS NEEDED BY WELL-PREPARED BEGINNING HIGH SCHOOL MATHEMATICS TEACHERS

This section provides additional detail, commentary, and examples of the knowledge, skills, and dispositions well-prepared high school mathematics teachers have, organized by the general standards described in Chapter 2.

STANDARD C.1. MATHEMATICS CONCEPTS, PRACTICES, AND CURRICULUM

Well-prepared beginning teachers of mathematics possess robust knowledge of mathematical and statistical concepts that underlie what they encounter in teaching. They engage in appropriate mathematical and statistical practices and support their students in doing the same. They can read, analyze, and discuss curriculum, assessment, and standards documents as well as students' mathematical productions.

Indicators include

C.1.1. Know Relevant Mathematical Content
C.1.2. Demonstrate Mathematical Practices and Processes
C.1.3. Exhibit Productive Mathematical Dispositions
C.1.4. Analyze the Mathematical Content of Curriculum
C.1.5. Analyze Mathematical Thinking
C.1.6. Use Mathematical Tools and Technology

Having strong subject-matter knowledge is critical for all well-prepared beginning teachers of mathematics, especially those at the high school level [C.1.1]. Intensive subject-matter preparation is necessary, given the level of the material; a more detailed outline of the essential mathematical understandings well-prepared beginning teachers of high school mathematics need is given below.

Many high school mathematics teacher candidates will have experienced success with a narrow school mathematics curriculum that did not promote conceptual knowledge or emphasize mathematical practices and processes. Thus, they must gain personal experiences with those practices and the ways they can support deeper knowledge of important mathematical concepts so that they can fully support their students' mathematical development [C.1.2]. For example, Vignette 7.1 demonstrates the close connection between the choice of a mathematical task and the teacher's role in facilitating discourse that is centered on mathematical ideas. This vignette is intended to illustrate how beginning teachers of mathematics at the high school level have deep knowledge of mathematics and skill in mathematical communication, reasoning, and sense making and move instruction beyond a focus on writing algebraic equations to making connections to other mathematical representations.

VIGNETTE 7.1. MEANINGFUL ALGEBRAIC EXPRESSIONS

The instructor of a methods class gave her students the following problem.

> *The height (in feet) of a ball t seconds after being thrown into the air is given by the following equation:*
>
> $$h(t) = 1\frac{3}{16} + 18t - 16t^2$$
>
> *What is the maximum height the ball will reach?* (Adapted from NCTM, 2009, pp. 32–33).

The instructor first asked the students to connect the function to models of projectile motion from physics, and the students discussed the meanings of the variables and coefficients in the equation. They also discussed how a function like this might be derived from data.

To answer the question, many students thought of taking the first derivative and solving for 0, knowing that the maximum height will occur at a critical point. While acknowledging that this approach is correct, the instructor urged her students to consider other ways that they could find the maximum height. While the students worked in small groups to explore the problem further, they devised a wide range of strategies including the following:

> Group 1 drew a graph of the function using a graphing utility and used it to estimate the maximum.
> Group 2 made a table of values to estimate the maximum. However, they further noticed that the function has zeroes at $t = \frac{19}{16}$ and $t = -\frac{1}{16}$. They reasoned that, because a parabola is symmetric, the maximum value will occur at the midpoint between those 2 zero values, $t = \frac{9}{16}$. So the maximum is $h\left(\frac{9}{16}\right) = \frac{100}{16}$.
>
> Group 3 students remembered that the method of completing the square could be used to find the vertex of a parabola, which they felt should be the maximum value. But they admitted that their understanding of completing the square was limited to having memorized a procedure, and they had forgotten the procedure.

In a class discussion, the instructor asked each group to present their solution to the class, carefully explaining their reasoning; the other students were encouraged to analyze the approach presented. After Group 1 presented its solution, another student asked, "Just because it looks like the maximum on the graph, are you sure that is the exact maximum height?" The Group 1 members admitted that they were approximating, so the value might not be exact. This exchange was an opportunity for the instructor to guide the discussion toward foundational aspects of rational and irrational numbers and precision, with the goal that the students articulate that the maximum might be irrational or it might be rational with a periodic decimal expansion that exceeds the numerical precision of graphing utility.

Group 2's solution was met with some amazement by the other students, who had not thought of taking such a simple approach. The instructor led a brief discussion concerning why the graph is symmetric.

Group 3 showed their progress but admitted that they were not entirely sure how to complete the square. The instructor allowed the other groups to discuss briefly among themselves, and the class reached agreement on the algebra—$h(t) = -16\left(t - \frac{9}{16}\right)^2 + \frac{100}{16}$. The instructor encouraged the students to discuss why this computation would reveal the maximum; that is, -16 times a square must be 0 or negative, so *h*(*t*) must be less than or equal to $\frac{100}{16}$. She also challenged the students to represent completing the square geometrically: "After all, doesn't it sound geometric?" She provided algebra tiles and asked the students to explore several simpler problems to better understand the method; she then asked them to use Desmos to represent this problem. She led a discussion of the underlying structure of the method of completing the square (cf. CBMS, 2012, p. 55). Finally, she explained that when they take their abstract algebra course the following semester, the students will explore connections between the complex numbers and the set of all polynomials in a variable *x* with real coefficients (cf. CBMS, 2012, p 59).

The instructor concluded the lesson by assigning the students to write a reflection on mathematics content and mathematical practices that they had investigated in the lesson.

Finally, high school mathematics teacher candidates need to gain productive dispositions toward engaging in mathematics, recognizing that even in the absence of a known solution method, they have resources that can help them make progress toward a solution. Despite challenges they may face in their advanced mathematics classes, they still see mathematics as an exciting and interesting endeavor [C.1.3]; without such dispositions, they will have little success in convincing their students of the value and importance of mathematics.

HS.1. Essential Understandings of Mathematics Concepts and Practices in High School Mathematics

> Well-prepared beginning teachers of mathematics at the high school level have solid and flexible knowledge of relevant mathematical concepts and procedures from the high school curriculum, including connections to material that comes before and after high school mathematics and the mathematical processes and practices in which their students will engage. Relevant mathematical concepts include algebra as generalized arithmetic, functions in mathematics, diagrams and definitions in geometry, and statistical models and statistical inference. [Elaboration of C.1.1 and C.1.2]

According to *The Mathematical Education of Teachers II (MET II)* (CBMS, 2012), well-prepared beginning teachers of mathematics at the high school level need to learn mathematical content that is "tailored to the work of teaching, examining connections between middle grades and high school mathematics as well as those between high school and college" (p. 54). In addition, they need to build strength with mathematical processes and practices; again according to *MET II*, they need a "full range of mathematical experience themselves: struggling with hard problems, discovering their own solutions, reasoning mathematically, modeling with mathematics, and developing mathematical habits of mind" (CBMS, 2012, p. 54) [see C.1.2]. As the Mathematical Association of America advocated for all who pursue mathematics majors, they must develop "effective thinking and communication skills" (Tucker, Burroughs, & Hodge, 2015, p. 1). Finally, they need productive dispositions toward mathematics, recognizing that mathematics is inherently a human activity [see C.1.3]. Many aspects of mathematics are rooted in the development of useful tools to address problems in the real world, and recognizing the impulse to find practical uses in mathematics provides both meaning and motivation to the study of mathematics. Acknowledging that mathematics has been created by a wide range of civilizations across history reinforces that it does not belong to a particular group.

The elaboration of mathematical content standards across this document relies on the expertise and thorough treatment of the topics as provided in the *MET II* report (CBMS, 2012). The structure of the *MET II* chapter related to high school mathematics differed from that of the other grade-band chapters in that it was organized around courses in a beginning teacher's undergraduate mathematics major that leads to teaching licensure, whereas the elementary and middle grades chapters were related to the content standards of the *Common Core State Standards – Mathematics* (*CCSS-M* [NGA & CCSSO, 2010]) as a framework. Thus, the structure of this chapter, relying on *MET II,* also differs from the structures of the early childhood, upper elementary, and middle level grade-band chapters in this volume. In the following, we provide four essential ideas about high school mathematics content for beginning teachers, including understanding algebra as generalized arithmetic, the role of functions in mathematics, the role of diagrams and definitions in geometry, and statistical models and statistical inference as well as the connections among these mathematical areas. While the high school mathematics curriculum continues to evolve, additional experiences in mathematical modeling and computer programming or coding may also be needed. Each of these areas is discussed below; the NCTM CAEP Mathematics Content for Secondary Addendum to the NCTM CAEP Standards 2012 (NCTM & CAEP, 2012b) has provided more detailed descriptions of the mathematical knowledge that beginning mathematics teachers should have and also reiterated the importance of mathematical processes and practices.

Reasoning in algebra. Well-prepared beginning teachers of mathematics at the high school level understand that students in the elementary grades focus on the meanings of numbers and operations on them as foundations for developing tools and techniques in arithmetic and that in the middle school grades, students move from this focus on arithmetic to a focus on algebra as generalized arithmetic (Usiskin, 2004). These teachers have strong understandings of the foundational concepts in Pre-K–8 number and operations and how they lead into algebraic thinking. Without a purposeful focus on this perspective, teachers can begin their careers relying on naive views of high school algebra as *symbol pushing*—rules performed on symbols without applying underlying reasoning. The *CCSS-M* mathematical practice "look for and express regularity in repeated reasoning" (NGA & CCSSO, 2010, p. 8) captures a broader purpose of high school algebra: to enable students to build expressions, functions, and equations to model situations. Embedded in an understanding of algebra is an understanding of the roles and nature of variables as a part of the language of mathematics. Students begin to develop this perspective at the middle level, and it is addressed in courses designed specifically for high school mathematics teacher candidates. This perspective is also appropriate in mathematics courses with broad audiences, from calculus, to introduction to proofs, to linear or abstract algebra.

Functions in mathematics. Well-prepared beginning teachers of mathematics at the high school level understand *big ideas* about functions (Lloyd & Beckmann-Kazez, 2010):

> The concept of function is intentionally broad and flexible.... Functions provide a means to describe how related quantities vary together.... Functions can be classified into different families, each with its own unique characteristics.... Functions can be combined by adding, subtracting, multiplying, dividing, and composing them... Functions can be represented in multiple ways, including algebraic, graphical, verbal, and tabular representations. (pp. 7–8)

These ideas have their foundations in middle-level mathematics, and well-prepared beginning teachers of high school mathematics understand the connections across grades. In addition to understanding the nature of functions, well-prepared beginning teachers of high school mathematics must have facility with the foundational functions that are studied in high school mathematics: linear, quadratic, and other polynomials; rational functions; exponential, logarithmic, and trigonometric functions; and functions that are recursively defined. This understanding of functions permeates the calculus sequence; although the tools of calculus are required of beginning teachers, it is not to be assumed that a preservice teacher who has completed the calculus sequence has investigated functions adequately to build an understanding of functions to teach them well. The functions of precalculus-level mathematics are then revisited in courses for middle and high school mathematics from an advanced perspective and are studied from a teaching perspective in methods courses.

Diagrams and definitions in geometry. Well-prepared beginning teachers of mathematics at the high school level understand that the role of geometry in high school differs from its role in middle school and that the objects of study in geometry in high school are "general properties of classes of figures" rather than "properties of individual figures" (Usiskin, 2004). They understand geometry from the perspective of transformations. Sinclair, Pimm, and Skelin (2012b) stated that "working with diagrams is central to geometric thinking" (p. 9). They further noted, "Geometry is about working with variance and invariance, despite appearing to be about theorems" (p. 22). "Working with and on definitions is central to geometry" (p. 36), and "a written proof is the endpoint of the process of proving" (p. 48). The ability to use dynamic geometry software to investigate and understand variance and invariance of geometric objects is essential.

Statistical models and statistical inference. Statistics holds a unique place in the high school mathematics curriculum. Statistics and mathematics as disciplines share a common foundation, but they are distinct areas, each with its own methods and traditions. The most notable distinction is that statistics is built on inference from inductive processes, whereas mathematics is decidedly deductive. However, statistics is generally included as part of the mathematics curriculum. Beginning teachers might easily, mistakenly believe that statistics is just another course in mathematics (like algebra, or geometry, or calculus) and undervalue the important distinctions between the two subjects.

The *Statistical Education of Teachers* (*SET*) (Franklin et al., 2015) described the problem-solving experiences and habits of mind that statisticians have identified as important to their discipline: A "modern data-analytic approach to statistical thinking, a simulation-based introduction to inference using appropriate technologies, and an introduction to formal inference" (p. 8) are the appropriate introductions to statistics for high school teacher preparation. The authors recommended attention to

> both randomization and classical procedures for comparing two parameters based on both independent and dependent samples (small and large), the basic principles of the design and analysis of sample surveys and experiments, inference in the simple linear regression model, and tests of independence/homogeneity for categorical data. (p. 8)

They further recommended a focus on statistical modeling "based on multiple regression techniques, including both categorical and numerical explanatory variables, exponential and power models (through data transformations), models for analyzing designed experiments, and logistic regression models" (p. 8). High school mathematics teachers hold the bulk of responsibility for ensuring the integrity of the statistics taught and learned in schools. Statistics, data analysis, and modeling offer students unique opportunities to address real-world problems that affect them in their communities.

Peck, Gould, and Miller (2013) identified five *big ideas* that are central to teaching high school statistics:

> Data consist of structure and variability.... Distributions describe variability.... Hypothesis tests answer the question 'do I think this could have happened by chance?'... The way in which data are collected matters.... Evaluating an estimator involves considering bias, precision, and the sampling method. (pp. 10–11)

This perspective on statistics and statistical reasoning is present in the statistics coursework of effective programs; a modeling course can address the distinctions between mathematical models and statistical models and offers an appropriate forum for discussing the distinction between mathematics and statistics to prepare teachers to address it in their high school classrooms.

New emphases. New emphases in the high school curriculum are part of the fluid nature of curriculum and its responses to the needs of society. NCTM (2016b) is embarking on a project to define with clarity and specificity pathways for high school mathematics. Well-prepared beginning teachers of mathematics at the high school level must be aware of changes to high school mathematics that are likely in the coming years.

One new emphasis is on mathematical modeling. Authors of *Guidelines for Assessment and Instruction in Mathematical Modeling Education* (Consortium for Mathematics and Its Applications [COMAP] & Society for Industrial and Applied Mathematics [SIAM], 2016) advocate the importance of mathematical modeling, "a process that uses mathematics to represent, analyze, make predictions or otherwise provide insight into real-world phenomena" (p. 8). Mathematical modeling is different from using manipulatives to "model mathematics" and from "direct modeling," both of which pertain to how mathematics is represented (Hirsch & Roth McDuffie, 2016, p. ix–x). Investigating mathematical concepts through engineering design or a basic modeling cycle provides not only opportunities to integrate other disciplines but also a means for students to improve their mathematical understandings (Usiskin, 2015). Well-prepared teachers have substantive experiences engaging in mathematical modeling so that they can understand mathematical modeling and its potential place in the curriculum.

A recent publication that considered the future of mathematics, *The Mathematical Sciences in 2025* (National Academies Press, 2013), states that "over the years there have been important shifts in the level of activity in certain subjects—for example, the growing significance of probabilistic methods, the rise of discrete mathematics, and the growing use of Bayesian statistics" (p. 72). The book identifies two new drivers of mathematics–computation and big data, and for both of these drivers it describes how discrete mathematics plays an important role. Well-prepared beginners should be aware of major concepts in discrete mathematics and recognize its future importance.

Finally, calls for increasing emphasis on computer science and coding are often delegated to mathematics teachers, given "similarities, connections, and intersections between the fields of computer science and mathematics" (NCTM, 2016a, p. 1). Indeed, writing computer code can be a powerful tool for solving mathematics problems, and computer coding can include interesting mathematics related to the design and analysis of algorithms—for example, understanding why the time required for a program designed to solve a problem can grow exponentially as the problem size increases, indicating that it may not be a fruitful approach. However, as acknowledged in NCTM's (2016a) position statement on *Computer Science and Mathematics Education*, well-prepared mathematics teachers must understand that computer science is not merely a subfield of mathematics but is a field in its own right that requires specialized knowledge to teach.

HS.2. Use of Tools and Technology to Teach High School Mathematics

> Well-prepared beginning teachers of mathematics at the high school level are proficient with tools and technology designed to support mathematical reasoning and sense making, both in doing mathematics themselves and in supporting student learning of mathematics. In particular, they develop expertise with spreadsheets, computer algebra systems, dynamic geometry software, statistical simulation and analysis software, and other mathematical action technologies as well as other tools, such as physical manipulatives. [Elaboration of C.1.6]

Use of technology is an expected part of society, and ability to use technology is expected in the workforce. Technology is also an important tool for doing mathematics, and well-prepared beginning teachers are proficient in its use. Technology is more than a computational aid. Well-prepared beginners are able to guide students in exploring how technology can be used to explore patterns, shape, transformations, and sequences. Technology can assist one in making connections between multiple representations, and it can help students communicate their mathematical ideas to their classmates. Well-prepared beginning teachers are particularly prepared to use "mathematical action technologies" (cf. NCTM, 2014, p. xi) useful for high school, including spreadsheets, dynamic geometry software, function-graphing utilities and graphing calculators; computer algebra systems; statistics simulation software; and other applets and technological tools that can enhance students' conceptual understanding of mathematical concepts. These are powerful tools for doing mathematics that will be a part of the lives of the students they teach.

In effective programs, use of these tools is embedded throughout candidates' preparation – in their content preparation, methods courses, and clinical experiences, including in courses *not* specifically designed for teachers, so that they can use them in meaningful ways. As stated in the *MET II* (CBMS, 2012),

> Teachers should become familiar with various software programs and technology platforms, learning how to use them to analyze data, to reduce computational overhead, to build computational models of mathematical objects, and to perform mathematical experiments. The experiences should include dynamic geometry environments, computer algebra systems, and statistical software, used both to apply what students know and as tools to help them understand new mathematical ideas—in college, and in high school. Not only can the proper use of technology make complex ideas tractable, it can also help one understand subtle mathematical concepts. At the same time, technology used in a superficial way, without connection to mathematical reasoning, can take up precious course time without advancing learning. (p. 57)

Well-prepared beginning high school mathematics teachers are comfortable using technology to engage in mathematics and to effectively support meaningful mathematics learning. They develop dispositions toward critically evaluating the appropriate use of emerging technologies and are prepared to respond to new technological tools when they become available.

On the one hand, although generic conveyance software such as PowerPoint and interactive whiteboards can provide valuable support for classroom instruction, such software is not inherently mathematical and does not provide the experiences essential for learning high school mathematics with technology. On the other hand, technology can provide powerful tools to enhance communication about mathematics within the classroom, including blogs and interactive platforms such as teacher.desmos.com. Effective beginning teachers view technology as an expected and vital part of the classroom, not as a replacement for effective teaching but as an embedded tool that students use to explore mathematics.

Additionally, well-prepared beginning high school teachers view online communications, such as the MathTwitterBlogosphere (#MTBoS), as an essential professional resource to increase their understanding of mathematics teaching, to share their triumphs and seek insights into their conundrums, and to build professional relationships with other mathematics teachers across the nation. At the same time, they maintain a critical ability to ascertain the value of content available on the Web and in social media and are prepared to respond to misinformation found there.

Effective beginning teachers are also prepared to use other nonelectronic tools such as manipulatives in their classrooms. Contrary to common beliefs, high school students benefit from using algebra tiles, 3D models, and other physical manipulatives (cf. NCTM, 2014). Well-prepared beginners have the knowledge needed to "make sound decisions about when such tools enhance teaching and learning, recognizing both the insights to be gained and possible limitations of such tools" (NCTM, 2012, p. 3).

STANDARD C.2. PEDAGOGICAL KNOWLEDGE AND PRACTICES FOR TEACHING MATHEMATICS

Well-prepared beginning teachers of mathematics have foundations of pedagogical knowledge, effective and equitable mathematics teaching practices, and positive and productive dispositions toward teaching mathematics to support students' sense making, understanding, and reasoning.

Indicators include

C.2.1. Promote Equitable Teaching
C.2.2. Plan for Effective Instruction
C.2.3. Implement Effective Instruction
C.2.4. Analyze Teaching Practice
C.2.5. Enhance Teaching Through Collaboration With Colleagues, Families, and Community Members

Candidates seeking to become high school mathematics teachers may come into programs with excitement about doing mathematics, but they may be less aware that many of the students they will teach may not share that excitement. Although their inclinations may be to focus on the students most like themselves, they need experiences that help them recognize that teaching is much more than sharing their mathematical knowledge with those few students who are already engaged in learning mathematics. They need to develop commitment to supporting learning by all students and to develop instructional practices that will help them achieve that commitment.

HS.3. Supporting Each and Every Student's Opportunity to Learn Mathematics

> Well-prepared beginning teachers of mathematics at the high school level understand the importance of providing each and every high school student with opportunities to learn mathematics that will enable him or her to think analytically and creatively in preparation for workforce, college, citizenship, and life. [Elaboration of C.2.1]

Well-prepared beginning high school mathematics teachers advocate for and ensure that each and every high school student is given the opportunity to learn mathematics in meaningful ways. In particular, well-prepared beginners give serious attention to students who are living in poverty, Latinx, Black, indigenous, and emergent multilinguals, students who have been historically excluded or marginalized in mathematics. All too often, whether intentional or not, secondary mathematics teachers become gatekeepers to students' opportunities to engage in coursework that will prepare them for a myriad of job opportunities and life events. On the one hand, effective teachers have high expectations for their students. They pose high-cognitive-demand tasks and maintain high levels of cognitive demand by questioning students in meaningful ways and facilitating discourse. Such practices support their students' development of relational understandings of mathematics (Skemp, 1976) and understandings of how one concept relates to different forms of mathematics, to themselves, and to the broader world. On the other hand, if a teacher creates an environment in which students engage mainly in memorizing rules and mimicking what the teacher does, then the students will not develop meaningful mathematical knowledge and skills needed for their future success, not only in continuing education or the workforce but also in understanding how mathematics can contribute solutions to broader issues facing society, such as poverty, cancer, or access to adequate housing. Well-prepared beginners have opportunities to understand that such topics are complex and to find ways to highlight that complexity through rigorous mathematical analysis.

Thus, well-prepared beginning high school teachers need to understand that their beliefs about what learning mathematics entails, along with their beliefs about students' cultural backgrounds, ability levels, gender identities, and other defining characteristics, affect how they interact with students. Such interactions largely determine how students will engage in mathematics and how much students will advance. Teachers' beliefs and school policies related to tracking have caused many students to be placed in courses in which the instruction they receive does not enable them to reason and make sense of mathematics; instead those students become less confident in themselves as learners and doers of mathematics and might not see the point of doing mathematics if it does not relate to their current or future lives. A number of scholars have shown that tracking leads to more unfavorable outcomes than positive ones (Horn, 2006; Oakes, 2008; Stiff & Johnson, 2011). Moreover, tracking or placement of students in particular courses is often based on demographic factors more than on students' knowledge and abilities (Oakes, 2008; Stiff & Johnson, 2011). When teachers of mathematics hold the powerful belief that all students, regardless of race, ethnicity, gender, socioeconomic status, language use, gender performance, immigrant status, or past performance in coursework, can and will succeed in learning important mathematics, they give their students advantages in learning mathematics that will prepare them for their futures.

Vignette 7.2 could be used with teacher candidates to discuss issues related to opportunities to learn. Mathematics teacher educators might ask teacher candidates the following questions to orchestrate a discussion focused on the vignette:

1. What might be some reasons for the teacher's taking a hands-off approach to some of the students on this particular day?
2. Why might some classrooms at the school have such small numbers of students?
3. What would have been a better approach if the teacher were trying to help the students to develop more autonomy?
4. What are some strategies that could have been used to ensure that each student was moving forward in his or her learning?

VIGNETTE 7.2. PROVIDING OPPORTUNITIES TO LEARN

This excerpt describes events that took place in a high school pre-algebra classroom.

> There were four students in the class today: Antonio, Vanessa, Alicia, and David. Three of the students are Black (Antonio, Vanessa, and Alicia) and one is White (David). When the class started, Vanessa asked me for help with her assignment. I assumed that the class would go as it had yesterday, with me working with Vanessa (because she was ahead of the others in the book) and Ms. Smith, the regular classroom teacher, working with the rest of the class. However, this is not what happened. I worked with Vanessa, but Ms. Smith did not work with the other students. David asked for and was given permission to go to the nurse's office. He was gone for most of the period. Antonio appeared to try a few times to get started on his assignment, but he seemed unsure how to begin. He asked Vanessa at least twice to come over to his desk and help him. But Ms. Smith refused to allow Vanessa to work with him. Antonio did not ask Ms. Smith for help, nor did she offer him any assistance. He appeared to get little, if any, work done during this class period. Similarly, Alicia appeared to be trying to answer the questions in the book, but, given what we have seen in the past, I have to wonder how much she understood of what she was doing. Like Antonio, she did not ask Ms. Smith for help, and no assistance was offered. Ms. Smith sat at her desk. She did not speak to Antonio or Alicia except to tell Antonio that he could not get help from Vanessa. At some point Ms. Smith began to make the model staircase that the students would be asked to construct in the next section of the book. When David returned to the classroom, he asked if he could also build a staircase. He worked on the model for the rest of the period. Thus, the class period ended with essentially no instructional interaction between teacher and students. After class, Ms. Smith indicated that she was intentionally taking a hands-off approach in which she would not give help until the students took the initiative to ask for it. This action (or, more accurately, inaction) appeared to be targeted at Antonio. Ms. Smith stated that he makes no effort and only disrupts the class. This appeared to be the justification for her hands-off approach.

We share this example because it was, in fact, representative of a larger pattern. What happened in this class period was not unique in our observations of the teachers' classes. The particular conditions of this class, specifically the small class size, highlight a phenomenon that could be seen elsewhere. We characterize this pattern of action and inaction as "allowing students to fail." As a result of Ms. Smith's hands-off approach, Antonio and Alicia, the only two students in the class in need of help, received no instruction from the teacher for an entire 50-minute class. Yet, this was but a starker example of events that took place in other teachers' classrooms as well. In other cases, students sat for the entire class period, essentially ignored by their teachers. These students were allowed to not work as long as they did not disrupt the class. Students who did not "take responsibility for their own learning" were subject to this hands-off approach.

Note. Adapted from "No Time Like the Present: Reflecting on Equity in School Mathematics" by C. Rousseau and W. F. Tate, 2003, *Theory into Practice*, 42(3), 210–216. Copyright 2003 by Taylor and Francis, Ltd. Adapted with permission.

Vignette 7.2 may cause some teacher candidates to gasp in disbelief and others to say that they have witnessed such situations in school placements or as students in high school. The discussion of the vignette or similar experiences will cause them to think about students' opportunities to learn and the role of teacher as an advocate for students, not simply a disseminator of information. Teachers need to engineer classrooms to provide students appropriate content exposure, content coverage, content emphasis, and quality instructional delivery (Tate, 2005). More than just planning for or implementing practices that support students to learn, mathematics teacher candidates need to look for opportunities to measure whether students would agree that they, indeed, have opportunities to learn.

STANDARD C.3. STUDENTS AS LEARNERS OF MATHEMATICS

Well-prepared beginning teachers of mathematics have foundational understandings of students' mathematical knowledge, skills, and dispositions. They also know how these understandings can contribute to effective teaching and are committed to expanding and deepening their knowledge of students as learners of mathematics.

Indicators include

- C.3.1. Anticipate and Attend to Students' Thinking About Mathematics Content
- C.3.2. Understand and Recognize Students' Engagement in Mathematical Practices
- C.3.3. Anticipate and Attend to Students' Mathematical Dispositions

Well-prepared beginning teachers of mathematics at the high school level recognize that their students may not think about mathematics the same way they do. They need to see their roles as building on the thinking of their students rather than only *transmitting* their mathematical knowledge to their students [C.3.1]. For example, when given a quadratic equation to solve, students may use a *guess-and-check* strategy. Rather than counting this strategy as wrong because it is not deductive in nature, well-prepared beginners consider how that solution could become part of the discourse around how to solve quadratic equations. Acknowledging the thinking behind that approach while also noting its limitations in situations in which solutions are not integers will motivate the need for other approaches. In this example, students are naturally engaged in the doing of mathematics, developing mathematical processes and practices [C.3.2].

By the time they are in high school, many students may conclude that they are not good at mathematics and certainly do not like studying it; well-prepared beginning teachers need to be cognizant of strategies for increasing their students' confidence in engaging mathematics as well as their appreciation for its usefulness [C.3.3].

STANDARD C.4. SOCIAL CONTEXTS OF MATHEMATICS TEACHING AND LEARNING

Well-prepared beginning teachers of mathematics realize that the social, historical, and institutional contexts of mathematics affect teaching and learning and know about and are committed to their critical roles as advocates for each and every student.

Indicators include

- C.4.1. Provide Access and Advancement
- C.4.2. Cultivate Positive Mathematical Identities
- C.4.3. Draw on Students' Mathematical Strengths
- C.4.4. Understand Power and Privilege in the History of Mathematics Education
- C.4.5. Enact Ethical Practice for Advocacy

Elaboration HS.3, Supporting the Opportunity to Learn by all High School Mathematics Teachers, states that well-prepared beginning high school teachers need to understand that their beliefs about what learning mathematics entails along with their beliefs related to students' cultural backgrounds, ability levels, and other defining characteristics affect how they interact with students. Such interactions largely determine how students engage in mathematics and how much students advance. In addition to elaborating Standard C.2 for the high school level, HS.3 also has significant implications for Standard C.4. The following elaboration further addresses the need to cultivate positive mathematical identities in each and every student.

HS.4. Cultivating Positive Mathematical Identities in Each and Every Student

> Well-prepared beginning teachers of mathematics at the high school level draw on students' strengths to cultivate positive mathematical identities. [Elaboration of C.4.2 and C. 4.3]

Within their programs, high school teacher candidates need to see diverse students reasoning and making sense of mathematics—students who differ in socioeconomic status, race/ethnicity, gender performance, immigrant status, religious background, ability, linguistic background, and in other ways. In a methods course, Vignette 7.3 can be used as a catalyst around developing positive student mathematics identities.

VIGNETTE 7.3. MATHEMATICAL IDENTITY

I am Michael Davis, an African American junior in a geometry class. I am motivated to do well in this class because I want to go to college, and if I make good grades, I might get a scholarship. I actually like this class because I am given opportunities to solve problems in groups with my peers. I like the discussions that we have, especially debates when we do not agree on a solution. I like activities that allow me to discover important connections between different topics of mathematics, between my life and mathematics, and between mathematics and what is going on in social media and the world. I also like making conjectures and testing them to find out if they are true.

Mathematics teacher candidates need opportunities to reflect on the identities that students develop and the factors that contribute to the contexts in which those identities arise. Questions the vignette could prompt include the following: Why might Michael feel the way he does about mathematics class? Given that Michael is a Black male, what might be noteworthy about his experience? Why might he enjoy working in a group? What does it take to develop meaningful *group-worthy* problems for students to solve? What would support the development of a classroom culture in which group work is valued? What are the differences between problems and exercises? Why might Michael care about relating mathematics to other topics, his life, social media, and the world? Should these connections be made between things with which he is already familiar, or can they also expand this knowledge? How does a teacher orchestrate this discussion well? In other words, what problems might be considered in this classroom beyond those teachers might consider real-world problems or curricular materials they might think include examples of real-world problems, but that a student sees as contrived or "not my real world?"

Furthermore, teacher candidates need opportunities within their programs to facilitate the mathematics learning of students from the aforementioned groups. Also, when possible, high school mathematics teacher candidates should collaborate with special-needs teachers and teacher candidates who are focusing on students with special needs. Mathematics teacher candidates need to learn to read Individual Education Plans (IEPs) for students and work with special-needs teachers to understand how to differentiate instruction, use multiple entry-level tasks, provide appropriate accommodations, and use other strategies to ensure that mathematics learning occurs in meaningful ways for each and every student.

High school teacher candidates need opportunities to interact with emerging multilingual students and to understand that just translating words for students is not enough; they need to create environments and learning situations in which the students can understand and own the mathematics for themselves. For example, students may be allowed to use algorithms from their home countries or to show their work in ways that are not standard practice in the United States (Civil & Menéndez, 2011; Civil & Planas, 2010; Gutiérrez, 2015).

High school teacher candidates need to understand how mathematics structures and vocabulary can challenge emergent multilingual students in multiple ways: Words have multiple meanings, different meanings in different contexts, and different pronunciations when used as noun versus as adjective. Thus, beginning high school teachers need to be aware of the ways in which language influences learning, including recognizing that all mathematics students are language learners (e.g., the word *function* does not mean the same thing to a monolingual English speaker when stated in the context of a broken washing machine at home versus in a mathematics classroom). Vignette 7.4 highlights a high school teacher candidate's revelations in learning to meet the needs of her emerging multilingual students.

This story illustrates the importance of not only having teacher candidates work with a diversity of learners but also having them reflect on and process those experiences to better understand how they can build on the cultural and linguistic resources that students bring to the classroom to support the learning of each and every student. This vignette may be used in a methods course to help teacher candidates consider what they would do if faced with a similar situation.

VIGNETTE 7.4. A STUDENT TEACHER'S REVELATION RELATED TO EMERGENT MULTILINGUAL STUDENTS

On the first day of school, my cooperating teacher learned that three Guatemalan students who did not speak English would join her 7th-period algebra course. I was immediately excited and eager to meet them. I have studied Spanish for many years and now have a bachelor's degree in the language, but I have had few opportunities to use my knowledge in the workplace. I knew that speaking Spanish would be useful for a teacher in the public school system, but I was surprised to get to use it so soon. Because of their language needs, this group of students has worked with me every day since Day 1. This experience has given me insight into the challenges that non-English-speaking students face daily at schools in which teachers have not been prepared to facilitate their learning and to help them achieve success.

During the initial observation weeks of student teaching, I spent my time in 7th period sitting with the Guatemalan students, providing explanations in Spanish of what transpired in class. In the first days of class, during prealgebra review, simple translation of the materials and interpretation of the teacher's lessons kept these students with the rest of the class. However, when the ideas discussed in class quickly became more complex, I realized that translations did not suffice. They understood the words I was saying, but not the mathematics. I realized then that these students, like any other students, needed instruction that they could understand. I began to shift my focus from translation and interpretation to teaching mathematics—a more difficult job that required much more thought and preparation. The same materials were not appropriate for both languages and both cultures, and I needed to differentiate my instruction. Because I was still observing, I was able to provide these materials and provide one-on-one assistance to these students.

While I continued to spend entire periods working with just the Guatemalan students, I wondered how they functioned in their other classes that had no Spanish-speakers to help them. I followed them through their schedule for a day to see what they were experiencing and found that they did nothing all day. They were ignored and not even given assignments in all but one class (Spanish class). I do not know how the teachers should have taught students with whom they could not communicate, but I think that doing nothing is the wrong solution.

When I began teaching 7th period myself, however, I realized that providing all the extra help needed by the Guatemalan students while teaching 25 other students in the classroom was very difficult (even though I spoke Spanish). My ability to help them diminished, and I started to see the real challenge in helping the struggling students without ignoring the other students.

I decided that to be successful, one must not separate the struggling students from the rest of the group. I began designing lessons to engage both groups of students and relied on multiple levels of knowledge to allow students at multiple levels to interact with one another. Additional assistance may be provided, but, in general, we must ensure that our lessons meet the needs of each and every student.

One fact stood out to me the more I worked with them: The Guatemalan students had a strong desire to participate and be a part of the class. I had seen just how little they were able to engage with anyone in their other classes, and I think they saw this mathematics class as one of their few opportunities to interact with other students and teachers. I saw that I could use this desire not only to motivate them to understand and ask questions but also to motivate other students to participate.

The Guatemalan students learned the numbers in English very quickly, and within a few weeks, they were calling out answers. Their English teacher facilitated this process by reinforcing use of words from other courses in her class. The students stopped saying numbers in Spanish and stated them only in English, even when we were talking privately. I helped them to create a mathematical dictionary in their notebooks, entering words like *equalities* and *inequalities;* in a short time, they began explaining their processes using these words. I was impressed by their fearlessness in class. Although they did not speak English well, they never hesitated to give solutions in front of the class. I believe that their willingness to share encouraged the other students in the class to speak out and share their own understandings.

One student seemed to have more difficulty than the other two. I learned from him and his Spanish teacher that he had not been to school in several years before coming to the United States. He was weak in mathematics but also not strong in reading, writing, or the Spanish language; his first language was Xinca, a language spoken by indigenous people in Guatemala. I saw that he needed more than the class could provide. After he did poorly on a test, I explained to him that to succeed in the class, he needed extra work. He agreed to work with me before school, and the next morning he was there.

We worked together one on one for 20 or 30 minutes each morning, and he eventually caught on to what we were doing in class. I found knowing when to keep him working on a topic and when to move on with the class challenging, but we tried to use the extra time as wisely as possible. After a few weeks, he came less often before school and became less dependent on me during class. He became more comfortable asking his peers questions and more confident on quizzes and tests. Eventually, he stopped coming in the mornings, and I felt confident that he would be able to work along with the other students in the classroom.

My experience working with this class and the Guatemalan students has taught me a lot about how to effectively support students who have unique language needs. I realized that the students had come into our mathematics classroom eager to participate and wanting to be part of the classroom. I simply needed to find the proper ways to support them in making sense of the U.S. conventions we use and to clarify the language when there were not easily identifiable cognates. I learned that instead of treating all the Guatemalan students as the same, I needed to probe the previous experiences they had in mathematics, how each one learned, what resources and expertise they had before entering my classroom, and what they needed from me to succeed. I have learned the importance of not lowering expectations for students based on their language needs, that they can achieve the same learning objectives in different ways. I have also learned to recognize when a student truly needs extra work, outside class time, to be able to function with peers. It is important to keep in mind that providing the same learning experiences for all students does not necessarily result in an environment that is equitable. All students should have the same opportunities to learn and succeed. Providing these opportunities may require adjustments or supplemental instruction, and incorporating these methods and the resources the students bring into a curriculum to take advantage of and celebrate the diversity of learners benefits not only the targeted students but also students from different backgrounds.

PART 2. ELABORATIONS OF THE CHARACTERISTICS NEEDED BY EFFECTIVE PROGRAMS PREPARING HIGH SCHOOL MATHEMATICS TEACHERS

This section provides additional detail, commentary, and examples of what preservice programs need to do to effectively prepare their students to teach high school mathematics. We begin with general discussion of the nature of high school mathematics teacher preparation programs. Further elaboration is provided, organized by the Standards in Chapter 3.

HS.5. Effective Programs to Support Preparation of Teachers of Mathematics at the High School Level

> Effective programs preparing teachers of mathematics at the high school level are specifically focused on preparing high school teachers of mathematics.

An effective program for preparing teachers of mathematics at the high school level provides coursework and experiences focused on the teaching of secondary mathematics. All too often, the prevailing attitude is that secondary mathematics teachers need only a degree in mathematics to be prepared to teach. Although having a strong content background is important, it is not sufficient (CBMS, 2012). Others may emphasize the need for general courses on pedagogy or education with related field experiences. Such a background has value, but it is also insufficient. A well-prepared beginning secondary teacher of mathematics needs coursework and experiences that build their understanding of both the content of, and teaching methods particular to, high school mathematics, given the social context in which mathematics teaching occurs.

An analysis of various degrees at a major state university revealed results about the preparation of teachers that were startling to us (Strutchens, 2012). Most degree programs required a significant number of hours specific to the major. For example, a degree in electrical engineering required 49 credit hours in coursework specific to electrical engineering, 31 credit hours of supporting content courses (mathematics and physics), and 15 credit hours in general principles of engineering. In contrast, a degree in secondary mathematics education at this university could include as few as 22 credit hours specific to mathematics education (which also includes a student-teaching experience), with 42 credit hours in mathematics courses taken by all mathematics majors and 15 credit hours in education courses taken by candidates from all teaching fields. Clearly, the balance of coursework in mathematics teacher preparation is out of alignment with other professional preparation programs. Students in secondary mathematics education programs need and deserve coursework to specifically prepare them for success in their field of study, including mathematics-specific methods courses and mathematics content courses specific to teaching as well as meaningful clinical experiences in secondary mathematics classrooms overseen by supervisors who have expertise in secondary mathematics across the grades they will be certified to teach; these courses are described in depth in the following sections. The example pathways in the final section of this chapter provide illustrations of how these requirements may be met in various types of programs preparing teachers of mathematics at the high school level.

High school teacher certification or licensure programs may include some or all of the middle grades; for example, a license may be valid for Grades 7–12, or 6–12, or even 5–12. Such programs must also meet the recommendations in this document for middle level preparation. A candidate prepared to teach high school is not automatically prepared to teach in the middle grades. Preparation for teaching mathematics at the middle level as addressed in Chapter 6 requires that well-prepared beginners have knowledge and skills about learners at this stage of their cognitive and affective development and about the intended curriculum at the middle grades that calls for integrative approaches and experience in teaching within interdisciplinary teams. Certainly, effective preparation for teaching high school mathematics provides many of the attributes needed to be effective with middle level learners, but the significant differences among learners at these different levels as well as the differences in how middle and high school curricula and schools are organized require that programs that prepare future teachers for broader grade ranges address standards in both this chapter and the previous one. One theme of this chapter, also reflected in *MET II* is "that the mathematical topics in courses for prospective high school teachers and in professional development for practicing teachers should be tailored to the work of teaching, examining connections between middle grades and high school mathematics as well as those between high school and college" (CBMS, 2012, p. 54).

This recommendation may present challenges for programs with small numbers of secondary mathematics education teacher candidates. Enrollment in such programs may be inadequate for them to offer coursework specific to mathematics education or to support clinical supervisors with mathematics-specific expertise to assist candidates at the high school level; offering such support for a broader range of grades is even more problematic. In the Mathematical Association of America's (MAA's) *Committee on the Undergraduate Program in Mathematics Guide to Majors in the Mathematical Sciences* (CUPM guide), Tucker, Burroughs, and Hodge (2015) provided options for such programs at the high school level, including "regional consortia, distance learning, or co-convening courses with in-service teachers seeking graduate credit or professional development" (p. 1). Alternatively, they might focus on providing outstanding preparation in mathematics, with the intent that students intending to pursue careers in teaching mathematics enroll in graduate programs to prepare to teach. Just as not every college or university has the capacity and resources needed to offer a degree in electrical engineering or other specialized majors, so it must be acknowledged that not every college or university has the capacity and resources needed to offer a degree in secondary mathematics education. **Institutions that cannot offer the necessary experiences should strive to develop the necessary partnerships or other approaches to meet both the standards in Chapters 2 and 3 and the elaborations in this chapter.** This recommendation may require some institutions to make difficult decisions about whether to offer a mathematics teacher preparation program at the high school level.

STANDARD P.1. PARTNERSHIPS

An effective mathematics teacher preparation program has significant input and participation from all appropriate stakeholders.

Indicators include

P.1.1. Engage All Partners Productively
P.1.2. Provide Institutional Support

Although partnerships are important at all levels of mathematics teacher preparation, they are particularly important in ensuring the effective preparation of teachers of mathematics at the high school level.

HS.6. Partnerships to Support Preparation of Teachers of Mathematics at the High School Level

> Effective programs preparing teachers of mathematics at the high school level engage mathematicians and statisticians, school partners, and other stakeholders in supporting the growth of their candidates to become well-prepared beginning teachers. [Elaboration of P.1.1]

Given that many candidates complete many more hours in mathematics content courses than in coursework particularly focused on mathematics teaching, mathematics department personnel must recognize and attend to the particular needs of secondary mathematics majors. Having close working relationships between mathematicians and mathematics educators helps to ensure that both courses specifically focused on mathematical content relevant to teaching and those for a wider range of majors include attention to the needs of high school mathematics majors. In addition, mentor teachers overseeing clinical experiences need to be included in discussions around high school mathematics teacher preparation to ensure alignment between the practices advocated within the preparation program and the instructional practices candidates both observe and are supported in implementing. Having close relationships among those responsible for the program and the mentor teachers beyond a sole focus on providing clinical experiences enhances the probability that this alignment will be achieved. Vignette 7.5 illustrates this point.

VIGNETTE 7.5. THE IMPORTANCE OF COHERENT AND CONSISTENT PROGRAM COMPONENTS

The prospective mathematics teachers enrolled in State University's one-year post-graduate secondary credential program love the two mathematics methods courses Robert teaches. He makes the course engaging yet relevant, and the students appreciate the innovative and empowering ways they are learning to think about mathematics, mathematics teaching, and students. State U. offers a fifth-year teacher credential program whereby students are enrolled in cohorts of 30 students, with each cohort led by a leader who is responsible for placing students into their student-teaching assignments. The students in each cohort together take a set of non-subject-specific courses, including social foundations, educational psychology, reading in the content area, teaching content to emergent multilinguals, and classroom management; once a week the secondary students from the different cohorts meet for their content-specific methods courses. Consequently, Robert teaches the mathematics methods courses to all the secondary-credential students. The students concurrently student teach, so Robert provides opportunities in his methods class for students to reflect on opportunities and challenges experienced by student teachers.

Over the years, one of Robert's major frustrations has been that student teachers generally do not observe the kinds of instruction from their mentor teachers that he has been promoting in his methods class. Robert has found that his message is often trumped by the classroom practice the candidates observe, and when Robert raised the issue with others in his department, the response was that student-teaching placements have always been made by the cohort leaders, who must consider such factors as the school's location, the principal's sense of the needs of his or her school, and convenience for student-teaching supervisors.

Lately, however, the situation dramatically improved when two colleagues, Lauren and Diane, were awarded a Noyce Master Teaching Fellows grant. Lauren and Diane recruited 40 exceptional secondary school mathematics and science teachers who began a 5-year professional development project focused on improving classroom instruction. Furthermore, because one goal of the Noyce Master Teaching Fellow Project is developing leadership, Lauren and Diane supported the Noyce teachers in emerging into leaders by helping them learn, for one thing, how to more effectively serve as mentor teachers for student teachers. After 2 years of the Noyce project, Lauren and Diane lobbied for and were permitted to create and lead a new mathematics/science, STEM-focused cohort for 30 of the secondary mathematics and science credential student applicants. As part of this new cohort, Lauren and Diane assumed the responsibility for assigning student teachers, and their highest priority was that the mentor teachers be selected on the basis of their excellent and innovative instruction. Furthermore, they arranged for many of the Noyce Master Teacher Fellows to serve as mentor teachers. For the first time, Robert has found that most of his mathematics methods students now see the kind of instruction in their assignments that he is encouraging in his mathematics methods class.

This vignette highlights the complexity of program effectiveness, which can be attained only through the coordination and shared goals among all those charged with teaching or supporting the candidates. Even the greatest mathematics methods class will have minimal effect on the students if the methods class stands alone, disconnected from other program components. Candidates must experience a coherent message among the many moving parts across the university and the local school partners. This example also highlights that commitment to a credential program is a necessary, but often insufficient, condition for program excellence; in this case, the impetus for program change was initiated by the professional activity of the faculty. In particular, funding of the Noyce grant set into motion a series of changes that led to an improved credential program. Without the symbiotic relationship among the mathematics educators' work inside and outside the credential program, the program improvement would not have occurred.

STANDARD P.2. OPPORTUNITIES TO LEARN MATHEMATICS

An effective mathematics teacher preparation program provides candidates with opportunities to learn mathematics and statistics that are purposefully focused on essential big ideas across content and processes that foster a coherent understanding of mathematics for teaching.

Indicators include

P.2.1. Attend to Mathematics Content Relevant to Teaching
P.2.2. Build Mathematical Practices and Processes
P.2.3. Provide Sustained Quality Experiences

In this section, we comment on how high school mathematics teacher preparation programs can support their candidates' opportunities to learn mathematics, particularly the specific content preparation they need.

HS.7. Mathematical Content Preparation of Teachers of Mathematics at the High School Level

> Effective programs preparing teachers of mathematics at the high school level are focused on the relevant content knowledge needed for teaching high school mathematics, including connections to material that comes before and after high school mathematics. Coursework consists of the equivalent of an undergraduate major in mathematics (including statistics) with at least three content courses particularly relevant to teaching high school mathematics and incorporating sufficient attention to a data-driven, simulation-based modeling approach to statistics. [Elaboration of P.2]

Table 3.2 summarizes the requirements for the content preparation of teachers of high school mathematics as follows:

- The equivalent of an undergraduate major in mathematics (including statistics) that includes three courses with a primary focus on high school mathematics from an advanced viewpoint.
- Three courses in statistics that include a data-analytic and simulation-based approach with a focus on statistical models and inference (to be incorporated in the undergraduate major).

These requirements build on the recommendations in Chapter 6 of *MET II* (CBMS, 2012), on the recommendations for high school programs in the CUPM guide (Tucker et al., 2015), and on *SET* (Franklin et al., 2015). We reiterate Tucker et al.'s point of view that a "traditional liberal arts major in mathematics is neither necessary nor sufficient preparation for teaching high school mathematics" (p. 1). Departments of mathematical sciences in effective programs preparing high school mathematics teachers commit the necessary resources to ensure that the mathematics and statistics courses in teacher preparation programs for high school mathematics address content that enables teacher candidates to focus on essential ideas of high school mathematics in an environment that engages them fully in the robust practice of mathematics.

In a mathematics-degree program that is tailored for prospective teachers, candidates need opportunities to examine the essential ideas of high school mathematics. Authors of the CUPM guide recommend lower division content in calculus, linear algebra, mathematical proof, and data-based statistics with a focus on statistical inference and upper division coursework in mathematics organized around a variety of topics. We reiterate the

CUPM's and *MET II's* main recommendation that the mathematics major for teachers include the equivalent of 9 semester-hours of coursework focused on "high school mathematics from an advanced standpoint" (CBMS, 2012, p. 55). The CUPM guide provides examples of organizing this coursework around geometry, algebra, analysis, modeling, number theory, discrete mathematics, or the history of mathematics; we recommend that programs include some attention to all these content areas, within courses designed specifically for teachers if possible. In these courses, the essential ideas of high school mathematics and their connections to middle grades and university mathematics are central. Chapter 6 of *MET II* provides a comprehensive analysis of such connections.

In addition to mathematics content, mathematical practices and processes form the foundation of mathematics coursework in effective programs. Program personnel attend to the increasing importance of statistics and ensure that prospective teachers engage with statistical ideas through three courses that follow the design principles articulated in *SET* (Franklin et al., 2015).

As a philosophy underlying the mathematical education of high school teachers, all mathematics courses that include prospective teachers (not just those specifically targeting prospective teachers) need to provide examples and opportunities for students to examine how related mathematical ideas exist within high school curricula. These examples and opportunities can be provided as naturally and automatically as examples are provided for prospective engineers, or nurses, or business executives, within most mathematics courses. In recognition that fitting appropriate experiences into existing programs can be challenging, many researchers are focusing on providing such examples. For instance, the MTE-Partnership's MODULE(S)^2 Research Action Cluster is developing modules related to key areas of mathematics content for teachers (e.g., geometry, statistics, and modeling). These unit-long modules can either be bundled into a domain-specific course (e.g., three modules on geometry) or into a *capstone* course including modules from across domains. Alternatively, they can be inserted into an existing mathematics course for students from a variety of majors, such as including one geometry module into a Euclidean geometry course. Some instructors have included one of these modules into a mathematics methods course. Collaborative efforts like this are helping to create resources to address the ongoing challenges of creating programs that provide high school mathematics teacher candidates with experiences in mathematics relevant to their chosen profession.

The varied states' requirements for teacher certification present ongoing challenges. Some states' requirements are coursework-based, others are content-exam-based (e.g., based on a minimum score on the Praxis II), and some are both. Some programs prepare teachers within a 120-credit 4-year degree that includes mathematics content, education coursework, and student teaching; others are fifth-year post-baccalaureate programs; and still others include a master's degree including initial teacher preparation. These differences preclude our recommending a one-size-fits-all program. In the examples at the end of this chapter, we describe sample high school teacher preparation programs from a variety of perspectives, fully recognizing that these illustrations do not completely capture the variation of the possibilities that exist. A particular challenge for graduate preparation programs is that prospective teachers may come to the program with undergraduate mathematics degrees designed to prepare them for graduate school in mathematics. Providing them with both additional mathematics content from the perspective of content knowledge relevant for teaching and education coursework within the confines of one year can prove challenging. However, programs need to include this essential background knowledge for well-prepared beginning high school mathematics teachers.

Finally, we reiterate our previous point that high school teacher certification or licensure programs that include some or all of the middle grades need to attend to the content needed for middle grades mathematics as described in Chapter 6.

STANDARD P.3. OPPORTUNITIES TO LEARN TO TEACH MATHEMATICS

An effective mathematics teacher preparation program provides candidates with multiple opportunities to learn to teach through mathematics-specific methods courses (or equivalent professional learning experiences) in which mathematics, practices for teaching mathematics, knowledge of students as learners, and the social contexts of mathematics teaching and learning are integrated.

Indicators include

P.3.1. Address Deep and Meaningful Mathematics Content Knowledge
P.3.2. Provide Foundations of Knowledge About Students as Mathematics Learners
P.3.3. Address the Social Contexts of Teaching and Learning
P.3.4. Incorporate Practice-Based Experiences
P.3.5. Provide Effective Mathematics Methods Instructors

In this section, we provide two elaborations describing how programs preparing teachers of mathematics at the high school level can support their candidates' opportunities to learn to effectively teach mathematics, Ethics and Values for Teaching Mathematics at the High School Level and Mathematics Methods Experiences for Teachers of High School Mathematics.

HS.8. Ethics and Values for Teaching Mathematics at the High School Level

> Effective programs preparing teachers of mathematics at the high school level provide multiple opportunities for candidates to develop political clarity on the profession and their advocacy roles in teaching. [Elaboration of P.3.3]

The important roles that teachers of high school mathematics play in teaching young adults during their final years of compulsory education and in potentially providing the bridge to further education, employment, and citizenship heighten the need for effective programs preparing teachers of mathematics at the high school level to provide opportunities for candidates to develop political clarity on the profession and their advocacy roles in teaching. Candidates must understand the possible influences of the curriculum and school policies on student learning and students' identities and be willing to advocate for their students when their students' best interests are being violated. Having *political clarity* requires that teachers ensure that the very students who have not been served well by the school system become the focus and the means by which we judge the efficacy of what we are doing in mathematics teaching, rather than considering the students as a single group.

In effective programs, mathematics methods experiences address the social, historical, political, and institutional contexts that affect mathematics teaching and learning and provide practice-based experiences to develop core practices and pedagogical content knowledge that honors mathematics, students' mathematical thinking, and cultural/community-based funds of knowledge and experiences. Instructors of mathematics methods experiences are aware of issues particular to high school students. For example, although the practice of sorting students into or away from high-level mathematics courses begins before high school, these tracks are often *set in stone* by high school, thereby precluding many students from accessing the most rigorous mathematics courses. Furthermore, although deficit language pervades all levels of schooling, deficit language is ingrained in the fabric of many mathematics departments. Moreover, many teachers may think that by high school age, those students who are *good* at mathematics have already been defined and little can be changed in students' dispositions toward the discipline or their motivations to learn. For example, teachers commonly gather in the lunchroom or mathematics office to complain about today's students' lack of skills (relative to students from the past) or to speak about groups of students as if they are a homogenous group, ignoring their individual differences and the unique strengths and experiences they bring to classrooms. Mathematics methods experiences in effective programs offer opportunities for candidates to learn to respond to and professionally challenge colleagues who may hold deficit perspectives on students.

Rather than simply learning about the latest reform initiatives in mathematics education, prospective teachers can benefit from opportunities to see that professionals do not always blindly follow recommendations of their district or state officials. Instead, candidates are exposed to both the need and specific strategies for creative insubordination, the bending of rules to adhere to higher ethical standards (Gutiérrez, 2013b, 2016) when working with colleagues. One way in which those preparing teachers of high school mathematics can accomplish this work is through offering opportunities for prospective teachers to analyze the context of mathematics teaching and learning and prepare for the political realities of the profession. That is, just as prospective teachers need time to rehearse for teaching through such activities as peer teaching or extensive lesson planning, they need opportunities to prepare for the kinds of controversial discussions and negotiations they will face as new teachers in a context of high-stakes education and in a country that has consistently failed to support Black, Latinx, American Indian, and other historically marginalized communities. If a candidate believes that a school policy or curricular initiative is unfair, he or she will need to practice challenging that policy, including knowing how to do such things as effectively *press for explanation*, *turn a rational issue into a moral one*, or *seek allies*. They must understand that a number of factors will help them be more strategic and effective. According to Gutiérrez (2016),

> Choosing an appropriate strategy requires we first recognize the kind of issue at stake (i.e., What power dynamics are operating? How does this issue relate to student learning and social justice?) and then consider the speaker(s), our relationship with them, and the context in which we find ourselves. (p. 54)

Given the growing knowledge base of both prospective and practicing teachers using creative insubordination, teacher education programs can draw on cases of effective teachers employing this practice.

Vignette 7.6 provides an example of the kinds of dilemmas and reflective exercises that secondary mathematics teacher education programs may provide as a means for beginning teachers to consider the ethical decisions they may face in their future teaching.

VIGNETTE 7.6. HELPING CANDIDATES CONSIDER ETHICAL DECISIONS THEY MAY FACE

Marcia is attending her mathematics department meeting. She likes her colleagues and generally agrees with their approaches to teaching. However, today, she learns that the district has decided to change their textbooks to a series that is text heavy and requires significantly more reading for students to understand the examples and begin the homework. She recognizes the need for occasional changes in texts to keep up with current reforms, but she is worried that the emergent multilingual students in her school may now require more of her attention because they may be less likely to use the text as a resource or they may be less likely to persevere in text-heavy word problems in homework sets. She asks herself, "Will students become confused with the extraneous words or unfamiliar contexts?" Many of the students she knows who speak more than one language have been transitioned out of multilingual education classes and may not appear to need special attention. She would like to raise this issue at her meeting, but she is in only her 2nd year of teaching and wonders if she should just stay quiet on policy issues until she has earned tenure. She hopes someone else in her department will speak up on behalf of the emergent multilingual students in their school. After some time in the meeting, however, she realizes that her colleagues do not recognize the potential hurdle these new texts may pose for students. Those preparing teachers of high school mathematics could ask teacher candidates to reflect on how they might handle the situation if they were in Marcia's shoes.

1. Suppose Marcia decides to speak up. How might she frame her concerns so that she does not appear to be simply complaining and has no solutions? What kind of language could she use in this public meeting that will promote dialogue and action rather than self-defense or resistance by her colleagues? What particular strategies for creative insubordination might she employ? What kinds of examples or information might elicit understanding and empathy from her colleagues or administrators rather than deficit perspectives toward students? How might she speak to individuals who do not share her expertise or commitment to emergent multilinguals? If, as a department, they decide to accept the district's choice, could Marcia convince the department to structure ways to support students with the new texts? What might she suggest? Can the old texts be retained as resources for students? If Marcia is unsatisfied with the outcomes of the meeting, can she engage her emergent multilingual students in ways that would not require knowledge by or consensus with other teachers?
2. Suppose Marcia decided not to speak up at the meeting but wanted to follow up later on. How might she decide where to direct her energies? Would raising the issue with the department chair be the right place to start? Or, should she begin with a sympathetic colleague who has more teaching experience or more years in the school? What role could data or research articles play in her attempts to convince her colleagues about a particular action? Could the students and parents play roles in helping her advocate for emergent multilingual students and others who may struggle with the new textbook? Might she find allies (e.g., nonmathematics teachers in the building) with more seniority? What collective effort might they make?

In such reflective exercises, prospective teachers might develop essays to provide *professional advice* to Marcia, produce a script, or role play to rehearse how Marcia might raise these issues and enlist the help of others in her mathematics department to advocate for the emergent multilinguals in the school. This practice of rehearsing through the activity of "In My Shoes" is being used in mathematics teacher education programs in various universities across the United States (see, e.g., Gutiérrez & Gregson, 2013; Gutiérrez, Gerardo, & Vargas, in press).

Finally, effective programs launch their students on trajectories of continued professional growth, providing candidates opportunities for building "their emerging leadership by being capable and effective advocates for excellence in mathematics teaching and learning and by holding themselves and their colleagues responsible for the mathematical success of all students" (MTE-Partnership, 2014, p. 5).

HS.9. Mathematics Methods Experiences for Teachers of Mathematics at the High School Level

> Effective programs preparing teachers of mathematics at the high school level provide candidates multiple opportunities to learn to teach mathematics effectively through the equivalent of three mathematics-specific methods courses. [Elaboration of P.3.4]

In effective programs preparing teachers of mathematics at the high school level, methods courses or equivalent experiences address deep and meaningful mathematics comprised of the interrelationship among concepts, procedures, reasoning and justification, problem solving, and development of productive dispositions (Kilpatrick, Swafford, & Findell, 2001). Effective programs include experiences where candidates revisit high school mathematical content from a rich and meaning-making perspective. Even having content courses focused on mathematics content relevant to teaching mathematics, as recommended in HS.7, will not, however, overcome the influence of a candidate's own K–12 learning experience if it was focused primarily on procedures. Furthermore, given the variety of paths taken by prospective teacher prior to credential programs, some students in mathematics methods courses will not have completed three especially designed mathematics courses for high school. Thus, mathematics methods courses must be focused not only on the methods of teaching mathematics but also on the *high school mathematics* for which those teaching methods are intended. In programs that offer fewer than three mathematics methods courses, experiences of integrating rich and deep mathematics with issues of teaching and learning must be included in other program components. In programs certifying mathematics teachers at both middle and high school, at least half of the content must be focused explicitly on high-school-level-mathematics topics.

In effective programs, teachers must also be prepared to hold strong foundations of knowledge about students as mathematics learners, including knowing about research-based progressions within well-defined content domains, understanding students' ways of engaging in mathematical practices, helping students share their mathematical thinking, and leveraging the diversity of students' thinking to advance instruction. Consequently, experiences for prospective high school teachers to investigate high school students' mathematical thinking are also incorporated into mathematics methods courses and early field experiences.

Mathematics methods experiences in effective programs also address the social, historical, political, and institutional contexts that affect mathematics teaching and learning and provide practice-based experiences to develop core practices and pedagogical content knowledge that honors mathematics, students' mathematical thinking, and cultural/community-based funds of knowledge and experience.

As important as it is for prospective teachers to grapple with issues of mathematics, students' thinking, and equity, the experience is more valuable for candidates when these issues arise simultaneously. How might mathematics methods instructors address issues of equity alongside issues of mathematical content and students' thinking? Consider an example by Civil (2016) of a prospective teacher who, after reading an article about algorithms taught in Latin American countries that differed from those taught in U.S. schools, said, "This is nice, but they need to learn to do things the U.S. way" (Civil, 2016, p. 220.) This case raises a myriad of issues, including valorization of knowledge, whereby the knowledge held by immigrant children is not valued as much as the knowledge that is taught in U.S. schools; the role of convention and principle in algorithms; and the great pedagogical value in understanding and building upon students' mathematical thinking. All these issues must be given space to arise and be discussed in mathematics methods classes.

STANDARD P.4. OPPORTUNITIES TO LEARN IN CLINICAL SETTINGS

An effective mathematics teacher preparation program includes clinical experiences that are guided on the basis of a shared vision of high-quality mathematics instruction and have sufficient support structures and personnel to provide coherent, developmentally appropriate opportunities for candidates to teach and to learn from their own teaching and the teaching of others.

Indicators include

P.4.1. Collaboratively Develop and Enact Clinical Experiences
P.4.2. Sequence School-Based Experiences
P.4.3. Provide Teaching Experiences With Diverse Learners
P.4.4. Recruit and Support Qualified Mentor Teachers and Supervisors

Clinical experiences are central to the preparation of teachers of mathematics at any level. But they are perhaps even more critical at the high school level than at other levels, given that many mathematics teaching candidates at the high school level may not have, as high school students, experienced mathematics instruction that promotes deep conceptual knowledge and development of mathematical practices and processes.

HS.10. Clinical Experiences for Teachers of Mathematics at the High School Level

> Effective programs preparing teachers of mathematics at the high school level provide clinical experiences in which candidates develop teaching practices that support the learning of conceptual knowledge and mathematical practices and processes for each and every student. [Elaboration of P.3.5]

Central to providing effective clinical experiences for mathematics teacher candidates at the high school level are mentor teachers who themselves meet the expectations discussed in Chapter 2. These mentor teachers support candidates in overcoming the possibly implicit belief that "teaching as they were taught" will be adequate for their students' learning because it (ostensibly) was adequate for theirs. Mentor teachers help candidates realize the limitations of typical instructional approaches in supporting the learning of rich mathematics by all students. They support the goals of the program to provide candidates continuity between experiences in university classrooms and their experiences in their clinical settings, so that candidates do not accept the dismissive mantra, "Don't listen to those people in the ivory tower."

Although many high school mathematics teacher candidates may have chosen their careers largely on the basis of their own enjoyment of mathematics, they need to build parallel enjoyment in helping all students (not just the students who already enjoy engaging in mathematics) to come to understand and appreciate mathematics. In effective programs, clinical experiences are carefully constructed to help candidates move beyond focusing on managing a classroom to truly engaging each and every student in doing mathematics.

In effective programs, candidates are supported in navigating the diverse cultures of high school mathematics teachers, noting that some teachers may initially be more committed to teaching mathematics than teaching children, and others may be only a few years older than the students they are teaching. Teachers engaged in improving their practice encounter the challenges of changing the very entrenched vision of *high school* into a more engaging, collaborative, interdisciplinary educational setting, while wrestling with traditional expectations of parents, policymakers, and other stakeholders about what high school mathematics should be. The inherent changes and diverse cultures require communities in which high school mathematics teachers can learn from one another and improve through collaboration.

Finally, candidates being prepared to teach a broader range of grades (e.g., 6–12 or 7–12) than high school need to have significant clinical experiences at both the middle and high school levels, given the differences between middle school and high school students. See Chapter 6 for further details on the clinical experiences needed by middle level candidates.

STANDARD P.5. RECRUITMENT AND RETENTION OF TEACHER CANDIDATES

An effective mathematics teacher preparation program attracts, nurtures, and graduates high-quality teachers of mathematics who are representative of diverse communities.

Indicators include

P.5.1. Recruit Strong Candidates
P.5.2. Address Diverse Community Needs
P.5.3. Provide Experiences and Support Structures

Given the critical shortage of well-prepared teachers of high school mathematics, for effective high school mathematics teacher preparation programs to devote significant attention to attracting prospective teachers who have been successful in mathematics is imperative. Recruitment can be particularly challenging, given the perception that teachers receive minimal pay relative to other professions that individuals with a baccalaureate in mathematics can pursue. It is also particularly hard to recruit individuals of color into teaching, again given the allure of higher paying and more prestigious jobs in fields that are also seeking to diversify their workplaces. The traditional pathway into mathematics teaching by those who love engaging in mathematics is still present, and effective programs capitalize on that pathway. But other pathways can be explored; for example, programs might appeal to those who find enjoyment in helping others, including career changers who are interested in contributing to society in positive ways.

Efforts must also be made to counter the narrative that high school teaching jobs are particularly low paying; some emerging research shows a significant difference between what STEM majors perceive to be teacher salaries and actual salaries. Although pay may be low, it is not as low as perceived by those who might consider a career as a mathematics teacher (Dickey, 2016). Effective programs spread the word that teaching mathematics at the high school level is a viable career option that can be very satisfying to those who pursue it.

CLOSING REMARKS

Well-prepared beginning teachers of mathematics at the high school level need knowledge, skills, and dispositions related to mathematics and statistics content relevant for teaching, along with processes and practices, mathematics-specific pedagogy, the needs of individual students and their cultural contexts, and the political context for teaching. Well-prepared beginners must understand the critical importance of providing each and every high school student with opportunities to learn mathematics that will effectively prepare them for their futures. Well-prepared beginners draw on students' strengths to cultivate positive mathematical identities that will contribute to their students' successful participation in mathematics.

To ensure candidates to teach mathematics at the high school level are indeed well prepared, effective programs need to specifically focus on preparing high school teachers of mathematics. Simply providing a mathematics degree without attending to the specific needs of candidates preparing to teach high school mathematics will not suffice. Effective programs include the equivalent of three content courses specifically designed for teachers of high school mathematics, three mathematics-specific methods courses (or equivalent experiences), and clinical experiences where candidates are mentored by high school mathematics teachers who themselves have the necessary knowledge, skills, and dispositions. Throughout these experiences, candidates should develop political clarity on the profession and their role as an advocate for their students and the profession. Partnerships with mathematicians and statisticians, school partners, and others involved in teacher preparation are essential in achieving the vision of this chapter.

POTENTIAL PATHWAYS FOR PREPARING HIGH SCHOOL MATHEMATICS TEACHERS

Programs to prepare high school mathematics teachers can be organized in many ways, depending on state policies, college or university guidelines, the intended audience (e.g., career changers), and so forth. However they are organized, effective programs must meet the requirements of the standards and their elaborations for high school in this document. Following are several examples of how differently organized programs can meet these recommendations.

Program 1. Four-year Bachelor of Science Degree

One approach to establishing a program that leads to the well-prepared beginning teacher of mathematics could be in the form of a four-year Bachelor of Science degree. This program has a major in mathematics with a teaching option. Consistent with *MET II* (CBMS, 2012), mathematics and statistics coursework consists of single and multivariable calculus, differential equations, two courses in data-based statistics and statistical inference, transition to proof, and linear algebra, each at the lower division. Upper division coursework includes three courses designed specifically for teachers, each of which allows the focus on essential ideas of high school mathematics: one course situates high school algebra and precalculus in the context of number theory, algebra, and analysis; a second course focuses on Euclidean and non-Euclidean geometry; and a third course focuses on modeling using the tools of mathematics without calculus and using simulation-based statistics. Each of these courses embeds the use of technology as a tool for learning mathematics. Mathematics coursework also includes three upper division elective courses in mathematics or statistics, and students choose from among courses including algebraic or geometric reasoning in the middle grades, history of mathematics, and others.

The program includes designated mathematics methods courses with field experiences, and embeds selected opportunities for working with students within mathematics content courses. Although the ideal program would include three methods courses, this program cannot currently do so, given the constraints of the program requirements as a whole. The program designers recognized this need for attention to teaching methods, and have addressed it by integrating pedagogy assignments and field experiences within mathematics content courses. The program includes education coursework completed by all high school teaching majors in the state, and includes student teaching.

Program 2. Integrated Fifth-Year Graduate Program

A second program achieves the recommendations of this chapter via a fifth-year graduate program that follows a strong undergraduate major in mathematics teaching. The undergraduate coursework includes three content courses addressing mathematical content for teaching. In addition to generic education coursework taken by all high school teaching majors, two content-specific high school mathematics methods courses, both of which include explicit attention to the teaching and learning of high school mathematics topics, are required, along with a course addressing using technology to teach mathematics taken by all middle and high school mathematics candidates. Furthermore, although equity issues are addressed in generic education courses, equity issues are also explicitly addressed within the mathematics methods courses, because only then will candidates understand that equity is not separate from, but fundamentally a part of, effectively teaching mathematics to high school students. Student teaching takes place concurrently during the methods courses, with the students assuming increased classroom-teaching responsibility over the course of each semester and from the first to the second semester. Because candidates student teach while taking courses, issues that arise in student teachers' practice are incorporated into the mathematics-methods-course discussions. Note that students coming from a general mathematics major that does not focus on mathematics teaching may have to complete pre- or co-requisite coursework.

Program 3. Graduate Program for Career Changers

An alternative version of the second program does not require prerequisite courses for mathematics teaching to be included in the undergraduate program, because the candidates may be career changers. Instead, this coursework is included in the program, embedded within courses addressing mathematical content relevant for teaching and mathematics-specific methods courses. Efforts are made to ensure that the program fully meets the standards and elaborations for high school mathematics preparation, noting that most candidates will require more than one year to complete the program. Depending on state requirements, candidates may be granted licensure or certification at the end of the first year and may then complete any remaining degree requirements the second year while employed as a full-time teacher.

Program 4. Liberal Arts Program

In a final program, the recommended standards are achieved in a liberal arts mathematics program that is a part of a coalition of universities and colleges offering high school mathematics teacher preparation. Specialized mathematics content courses for teachers as well as mathematics-specific methods courses are offered collaboratively with a nearby state university that leads a local partnership of higher education institutions and school districts. Some of these jointly offered courses are delivered online. In other cases, students meet on one of the campuses or at a central location. In this model, students in the smaller program can complete the requirements for becoming high school mathematics teachers that meets the recommendations in this document, and students at the state university can interact with colleagues from another context, thus broadening their awareness of issues related to mathematics education, particularly if the liberal arts college reflects demographics different from those of the state university.

CHAPTER 8. ASSESSING MATHEMATICS TEACHER PREPARATION

Assessment is a cornerstone for all educational endeavors, the means by which those striving to support learning come to know the knowledge, skills, and dispositions of those whose learning they are supporting. This information is crucial for a myriad of reasons, including, but not limited to the design of learning experiences, the development of curriculum and programs, and reflections of learners and their teachers. Assessment in mathematics teacher preparation is purpose driven, and those purposes are summarized in Table 8.1.

TABLE 8.1. PURPOSES FOR ASSESSMENT OF MATHEMATICS TEACHER PREPARATION

	Formative Assessment	Summative Assessment
Assessment of Candidates	Provides information on candidates' progress toward becoming well-prepared beginning teachers of mathematics	Determines candidates' readiness to begin practice as beginning teachers of mathematics (credentialing or licensing)
Assessment of Programs	Elucidates program's strengths and weaknesses to improve the program	Determines whether standards are met for program accreditation

These purposes are served through sound and efficient assessments designed to provide information, support interpretation, and spur action. Assessments necessarily appraise mathematics teacher candidates and elements of the mathematics teacher preparation programs that support their professional growth, providing timely formative information needed to provide ongoing feedback to candidates and to tailor learning experiences that support their growth. Assessments also provide summative information on the outcomes of teacher preparation, most crucially providing information on the quality of the knowledge, skills, and dispositions of mathematics teaching candidates and the candidates' proficiencies in supporting mathematical learning as well as the quality of a program.

AMTE is committed to promoting the improvement of mathematics teacher preparation. **Therefore, this chapter is focused on formative, improvement-oriented uses of assessment that are aligned with the teacher preparation standards described in earlier chapters.** Chapters 2 and 3 of this document set forth standards for what candidates should know and be able to do (Candidate Standards) as well as what programs must provide to support their candidates in achieving those standards (Program Standards). Effective programs need assessments that provide useful information about their candidates, both individually and collectively, to identify areas in which programs can better support their candidates' growth toward becoming well-prepared beginning teachers of mathematics.

Although we choose to focus on assessing these standards for purposes of ongoing improvement, we acknowledge that the standards could at some points and in some ways inform summative purposes. For example, these standards are not designed to establish minimum benchmarks that candidates must meet before beginning their careers as mathematics teachers. However, these standards provide important descriptions of knowledge, skills, and dispositions to be considered in devising such benchmarks. Such an emphasis on program quality and improvement is consistent with Standard 5 of the accreditation standards put forward by the Council for the Accreditation of Educator Preparation (CAEP, 2013b).

This chapter begins with recommendations about general features of effective assessments used in mathematics teacher preparation; these recommendations are abbreviated AF (Assessment Features). The recommendations for features of effective assessments should guide assessment decisions of instructors and programs when they select, design, and use assessments.

The second section of the chapter includes recommendations for assessments designed to determine progress toward attaining the Candidate Standards in Chapter 2; these recommendations are abbreviated AC (Assessment of Candidates). Finally, recommendations are given for assessments designed to determine progress toward attaining the Program Standards in Chapter 3; these recommendations are abbreviated AP (Assessment of Programs). Together these recommendations should be used to effectively guide efforts to support candidate growth and improve mathematics teacher preparation programs. A summary of the recommendations follows in Table 8.2.

TABLE 8.2. RECOMMENDATIONS ABOUT ASSESSING MATHEMATICS TEACHER PREPARATION

Features of Effective Assessments Used in Mathematics Teacher Preparation

Recommendation AF.1. Focus on Mathematics Teaching in Assessments

Effective assessments are tailored to provide information relevant to the development of mathematics teacher candidates.

Recommendation AF.2. Promote Equity and Access in Assessments

Effective assessments are designed, implemented, interpreted, and used to advance equity and access.

Recommendation AF.3. Embody Openness in Assessments

Effective assessments are part of a transparent process in which those assessing, those being assessed, and other stakeholders know the assessment focus, the nature of assessment methods and criteria, and the uses of the information gathered.

Recommendation AF.4: Support Valid Inferences and Action Based on Assessments

Effective assessments appropriately include multiple measures that are well suited to gather the types and amounts of information needed to make valid inferences and take action.

Recommendation AF.5: Embody Coherence and Sustainability in Assessments

Effective assessments are aligned with teacher preparation goals, opportunities to learn, and intended uses of the information they generate. They are designed and enacted in ways that can be sustained over time.

Assessing Quality of Mathematics Teacher Candidates

Recommendation AC.1. Assessment of Mathematical Knowledge Relevant to Teaching

Effective assessments of mathematics knowledge incorporate attention to candidates' development of mathematical knowledge relevant to teaching, including processes and practices.

Recommendation AC.2. Assessment of Mathematics Teaching Practice

Effective assessments of mathematics teaching practice include observations of teaching focused on how well the teaching supports learning of important mathematical content, processes, and practices by each and every student.

Recommendation AC.3. Assessment of Dispositions

Effective assessments provide data on a range of dispositions related to mathematics teaching, including dispositions toward engaging in mathematics, identity as a mathematics teacher and learner, and commitment to support the mathematics learning of each and every student.

Assessing Quality of Mathematics Teacher Preparation Programs

Recommendation AP.1. Assessment of Stakeholder Engagement

Effective mathematics teacher preparation programs assess the degree to which a broad range of stakeholders are meaningfully engaged in the program and involved in its ongoing improvement.

Recommendation AP.2. Assessment of Program Curriculum and Instruction

Effective mathematics teacher preparation programs assess the effectiveness of courses or equivalent professional-learning opportunities they offer in promoting their candidates' progress and work to improve them.

Recommendation AP.3. Assessment of Effective Clinical Experiences

Effective mathematics teacher preparation programs assess and work to improve the quality of the clinical experiences they provide mathematics teacher candidates.

Recommendation AP.4. Assessment of Recruitment and Retention

Effective mathematics teacher preparation programs assess and work to improve their success in recruiting prospective teachers, in retaining candidates in the program, and in supporting beginning teachers' continued professional growth, with a focus on developing a mathematics teacher workforce that reflects the diversity of the student population in the region.

Note that the match between these recommendations for assessment and the Candidate and Program Standards in Chapters 2 and 3 is not one to one. The features of effective assessments apply to all assessments of both candidate and program quality. Table 8.3 illustrates the relationships between the recommendations for assessing mathematics teacher candidate quality and the Candidate Standards; the checked cells indicate that the recommendation should particularly inform assessment of the given standard.

TABLE 8.3. RELATION OF ASSESSMENT RECOMMENDATIONS TO THE CANDIDATE STANDARDS

	Candidate Standards			
Assessment Recommendations	C.1. Mathematics Concepts, Practices, and Curriculum	C.2. Pedagogical Knowledge and Practices for Teaching Mathematics	C.3. Students as Learners of Mathematics	C.4. Social Contexts of Mathematics Teaching and Learning
AC.1. Assessment of Mathematical Knowledge Relevant to Teaching	✔			
AC.2. Assessment of Mathematics Teaching Practice		✔	✔	✔
AC.3. Assessment of Dispositions	✔	✔	✔	✔

Table 8.4 illustrates the relationships between the recommendations for assessing program quality and the Program Standards; checked cells indicate the recommendation should particularly inform assessment of the given standard. In addition, given that candidate outcomes are essential components of program success, the recommendations for assessing mathematics teacher quality also inform assessment of Program Standards.

TABLE 8.4. RELATION OF ASSESSMENT RECOMMENDATIONS TO PROGRAM STANDARDS

	Program Standards				
Assessment Recommendations	P.1. Partnerships	P.2. Opportunities to Learn Mathematics	P.3. Opportunities to Learn to Teach Mathematics	P.4. Opportunities to Learn in Clinical Settings	P.5. Recruitment and Retention of Teacher Candidates
AP.1. Assessment of Stakeholder Engagement	✔				
AP.2. Assessment of Program Curriculum and Instruction		✔	✔		
AP.3. Assessment of Effective Clinical Experiences				✔	
AP.4. Assessment of Recruitment and Retention					✔
AC.1. Assessment of Mathematical Knowledge Relevant to Teaching		✔		✔	
AC.2. Assessment of Mathematics Teaching Practice			✔	✔	
AC.3. Assessment of Dispositions		✔	✔	✔	

FEATURES OF EFFECTIVE ASSESSMENTS USED IN MATHEMATICS TEACHER PREPARATION

Research and professional standards provide ample guidance about the traits of sound assessments. In the case of assessments that will be generative for mathematics teacher preparation, literature and standards focused on teacher education and mathematics education are certainly useful. One source that was particularly pivotal in our consideration of the recommendations below was the National Council of Teachers of Mathematics *Assessment Standards for School Mathematics* (NCTM, 1995). Although the standards presented in that document were focused on assessment in Pre-K–12 mathematics 20 years ago, the ways in which they identify goals and traits of high-quality approaches to assessment make them both timely and applicable to assessment in mathematics teacher preparation.

The recommendations on the features of assessments provide a set of criteria by which individual assessments and systems of assessments can be judged. More generally, they relate to the content and forms of assessment as well as to routines that support the implementation of assessments and uses of assessment information. The recommendations can be used proactively as goals for assessment selection and development and can also shape dispositions guiding implementation of assessments.

Recommendation AF.1. Focus on Mathematics Teaching in Assessments

Effective assessments are tailored to provide information relevant to the development of mathematics teacher candidates.

Effective assessments in mathematics teacher preparation address key knowledge, skills, and dispositions that teacher candidates need to support mathematics learning. Effective assessments address candidates' understanding of mathematical concepts and their skills in using mathematics to solve problems. At the program level, effective assessments facilitate the capture and aggregation of information about mathematics teaching. Two components of assessment practice are entailed in the focus on mathematics teaching. First, assessments must be sensitive to the *mathematical* aspects of teaching, as opposed to relying on generic or subject-matter-neutral approaches. Second, although programs can assess aspects relevant to mathematics teaching outside of practice, assessments must appraise candidates' *engagement in mathematics teaching,* including supporting candidates' self-assessment of their own mathematics teaching. Given the complexity of mathematics teaching, those preparing teachers of mathematics often engage candidates in approximations of teaching in which some aspects of teaching are authentic but other aspects are temporarily suspended. Assessments can and should be used in these contexts to provide information on candidates' growth that can be used to shape candidates' subsequent opportunities to learn about mathematics teaching.

Recommendation AF.2. Promote Equity and Access in Assessments

Effective assessments are designed, implemented, interpreted, and used to advance equity and access.

Effective assessments in mathematics teacher preparation allow all candidates equitable opportunities to demonstrate their knowledge, skills, and dispositions. This process includes access for each and every candidate to assessments through which they can demonstrate their abilities to teach mathematics and to receive useful, mathematics-teaching-focused feedback. Additionally, criteria defining mathematics teaching proficiency honor diverse approaches in the service of students' meaningful and productive engagement with mathematics. All assessments are routinely analyzed for signs of bias or patterns of outcomes that indicate that the assessment is functioning inequitably. Results of assessments should support candidates' agency, leading to improvement of their opportunities to learn about mathematics teaching, including about differentiation of supports and resources to enhance the performance of each and every candidate. The tools used should produce reliable results and be valid in what they measure.

Recommendation AF.3. Embody Openness in Assessments

Effective assessments are part of a transparent process in which those assessing, those being assessed, and other stakeholders know the assessment focus, the nature of assessment methods and criteria, and the uses of the information gathered.

Transparency in assessment means that all involved in assessments and those stakeholders who use information from assessments are fully informed and have suitable roles in assessment selection, implementation, and use of assessment results. The goals for assessments must be clear to all involved, and those involved must validate that assessments provide useful and actionable information. That the assessment process be open to mathematics teacher candidates is particularly important. This openness can be attained in many ways, for example, by including self-assessment components in assessments and involving candidates in the development of assessment criteria. Candidates and other users of assessment results receive performance information in timely ways and in forms that are easily interpreted and meaningful. Although at times some aspects of assessments need to be secure, effective mathematics teacher preparation programs maximize opportunities to enhance the openness of the process and content of assessments.

Recommendation AF.4. Support Valid Inferences and Action Based on Assessments

Effective assessments appropriately include multiple measures that are well suited to gather the types and amounts of information needed to make valid inferences and take action.

Assessments of mathematics candidate and program quality should be comprehensive, utilizing multiple measures to address the range of relevant standards. Relying on a single data point in required assessments (such as a score on a test required for teacher certification) will not provide the breadth of information that is needed. Effective assessments must specifically address mathematics teaching beyond general measures that provide inadequate information about candidates' readiness to teach mathematics. Effective assessments extend beyond a sole focus on candidates' mathematical knowledge to include attention to their instructional practices and dispositions related to mathematics, as outlined in Chapter 2. The set of assessments chosen could include those that are locally designed and others that are selected from assessments that have been developed and used by the wider educational community. Users of assessments must be aware of the limitations of each assessment that is chosen and account for those limitations through warranted modifications or specific approaches to implementation, analysis, and interpretation of assessment information.

Assessment approaches should be chosen for their capacities to capture information most relevant to the knowledge, skills, and dispositions targeted in the assessment. Ease of administration is not the leading consideration in choosing assessment methods, just as assessment targets should be chosen by their importance rather than the ease of assessing them. Further, many valued outcomes are not easily or well conveyed quantitatively. Questionnaires, interviews, focus groups, observational notes, video recordings of teaching, and other records of practice provide critical complementary data. Although the learning of mathematics achieved by their students may be the ultimate assessment of a candidate's quality, measures of student success that focus on a single high-stakes assessment are unlikely to be valid or to provide information that can guide improvement.

Table 8.5 provides an overview of potential attributes and measures of interest to mathematics teacher educators. A collection of measures that provide a program the most meaningful data regarding their candidates' mathematical knowledge, instructional practice, and dispositions should be selected, judged on the basis of the values and goals of the program.

TABLE 8.5. MATHEMATICS TEACHER PREPARATION QUALITY: ATTRIBUTES AND EVIDENCE

Attributes	Possible Evidence
Program admissions and recruitment criteria (Potentially relevant for AC.1, AC.3, and AP.4)	• GPA • Average entrance-exam scores (e.g., basic-skills exam, state-required multiple-subject or mathematics exams) • Dispositions self-assessment and feedback from recommenders • Number and diversity of candidates
Quality and substance of instruction (Potentially relevant for AC.1 and AP.2)	• Syllabi, instructional materials, learning tasks for courses (or equivalent professional-learning opportunities) • Requirements for courses (or equivalent professional-learning opportunities) including those addressing mathematics • Common assessments for courses, including common rubrics measuring both mathematical content knowledge and ability to engage in mathematical practices measuring candidate progress and growth over time • Instructor use of active-learning and inquiry-based strategies • Observations/video recordings to support instructor reflection
Quality of clinical experiences (Potentially relevant for AC.1, AC.2, AC.3, AP.2, and AP.3)	• Fieldwork policies, including required hours • Qualifications of fieldwork mentors • Surveys of candidates' perceptions about quality of clinical experience • Field evaluations that are aligned with effective mathematics teaching practices • Data regarding candidate effect on student understanding and achievement (e.g., equitable student learning across various student groups) • Observations/video recordings to support candidate reflection • Student teaching logs documenting collaboration with mentors, field supervisors, and other building colleagues • Dispositions self-assessment and feedback from mentors and field supervisors
Faculty/instructor quality (Potentially relevant for AP.2 and AP.3)	• Percentage of faculty with advanced degrees • Percentage of faculty that are full-time, part-time, adjunct • Percentage of faculty with relevant Pre-K–12 teaching experience • Percentage of faculty with recent Pre-K–12 teaching experience • Surveys of faculty effectiveness
Effectiveness in preparing new teachers who are employable and stay in the field (Potentially relevant for AC.1, AC.2, and AP.4)	• Exit and alumni survey on preparedness • Ratings from program's field evaluation • Teacher-performance assessments • Hiring and retention rates • Ratings of graduates by principals/employers

Note. Adapted from *Evaluation of Teacher Preparation Programs: Purposes, Methods, and Policy Options* (p. 27), by M. J. Feuer, R. E. Floden, N. Chudowsky, and J. Ahn, 2013, Washington, DC: National Academy of Education. Copyright 2013 by National Academy of Education.

These examples are not comprehensive. Moreover, do not conclude from this list that the focus should be on the number of assessments used. Rather, the examples convey an array of assessment data that could be relevant, contexts in which assessments may occur, and people who could beneficially be involved in such assessments.

Finally, although many assessments related to mathematics teacher preparation may be mandated by university or state policy, significant benefits may ensue from including measures used across many universities. First, such common measures can be designed to align to the standards in this document, utilizing expertise across universities to provide economy of scale in their development, validation, and perhaps even scoring. Second, having common measures allows for comparisons across programs to better assess the progress being made in particular areas. For example, the Mathematics Teacher Education Partnership, a collaboration of secondary mathematics teacher preparation programs, has adopted several measures that include a common classroom-observation protocol (Gleason, Livers, & Zelkowski, 2015), a Partnership-developed survey in which candidates self-assess their readiness to teach, a Partnership-developed program self-assessment of progress along various dimensions, and a survey of the number of candidates produced by programs. These common measures are incorporated into each program's existing assessment system to provide information about the progress of the Partnership as a whole as well as the progress of individual programs in comparison to progress of the Partnership as a whole (Martin, W. G., & Gobstein, 2015). Such work also provides a foundation for collaboration on ways of using information provided by assessments and approaches to enhance teacher preparation in areas for which assessments across institutions show similar performance (joint work to design new efforts) or difference in performance (one institution with an area of strength could help partner institutions at which performance is less robust).

Recommendation AF.5. Embody Coherence and Sustainability in Assessments

Effective assessments are aligned with teacher preparation goals, opportunities to learn, and intended uses of information they generate. They are designed and enacted in ways that can be sustained over time.

In effective mathematics teacher preparation programs, assessments are logical outgrowths of program goals and learning experiences. This coherence makes the content within assessments and approaches to assessment seem natural and expected—instead of surprising, unfair, or alarming. Further, the strong connection also positions users of assessment data to see the ways in which insights relate to learning opportunities and project into actions.

Coherence across assessments includes coherence of the content assessed and the implementation and interpretation of assessments. Those implementing assessments do so in accordance with the specific requirements specified for each assessment. In addition, they have routines and norms that can be used across assessments to ensure appropriate and purposeful engagement as well as to mitigate sources of bias and misuse of assessment information. Formative and summative assessments complement each other and collectively provide multiple points of data on candidate and program outcomes.

Assessments in mathematics teacher preparation need to be chosen, enacted, and used in ways that can be replicated over time, not simply at a particular point in time. Routinely, efforts are made to enhance sustainability of the assessment system. However these efforts do not lead to assessments within the system being lightly or frequently changed. Attending to sustainability in these ways supports the collection of data necessary to answer important longitudinal questions about course and program effectiveness; the nature and degree of consensus among and between candidates, mentor teachers, and program instructors with respect to mathematics teaching quality; and the correlation, or at least correspondence, between particular assessments used by programs.

ASSESSING QUALITY OF MATHEMATICS TEACHER CANDIDATES

In effective mathematics teacher preparation programs, assessments of the quality of mathematics teacher candidates are selected or developed to provide insight into the knowledge, skills, and dispositions of the candidates. Providing such insight requires that assessments are sensitive to the mathematical and pedagogical dimensions of quality mathematics teaching. The features of effective assessments discussed in the first section guide the selection and use of assessments in an effective program. Effective assessments are accessible and promote equity, both in the sense of equitably appraising the mathematics teacher candidates themselves and assessing the ability of candidates to engage in teaching practices that promote the mathematical achievement of all their students. Every mathematics teacher candidate is fully aware of the assessments, products, criteria, and potential consequences of the assessments in which they engage. Those assessing the quality of mathematics teacher candidates draw on assessments and develop systems that are valid, coherent, and sustainable.

Assessments of candidate teachers' quality can be used in several ways to improve mathematics teacher preparation. Candidates can be positioned to learn from video and other records of their own teaching experiences and to apply criteria reflectively to their performance. Mentor teachers and field supervisors can analyze appraisals of teacher candidates' teaching with an eye toward providing opportunities to further hone particular teaching practices or to expand knowledge of children's thinking and learning. Course instructors can look across reflections of teacher candidates to notice areas in which they could raise awareness of bias that might be interfering with their abilities to formulate productive next steps in teaching. Content course instructors could look at student work that is collected by teacher candidates to prioritize which mathematical understandings, principles, or practices to bring to the fore. Program administrators could look at the performance of teacher candidates at a particular point in the program and marshal resources that could expand or deepen work on a particular teaching practice.

This section provides recommendations for how effective assessments can be used to gather information about mathematics teacher candidate quality to support candidate growth toward meeting the standards from Chapter 2, as outlined in Table 8.3.

Recommendation AC.1. Assessment of Mathematical Knowledge Relevant to Teaching

Effective assessments of mathematics knowledge incorporate attention to candidates' development of mathematical knowledge relevant to teaching, including processes and practices.

Effective mathematics teacher preparation programs include a variety of assessment methods to assess the (a) mathematical and statistical content knowledge including both conceptual and procedural knowledge, and (b) knowledge and use of mathematical and statistical practices and processes of teacher candidates. Effective programs use more than tests embedded in coursework or a mandated, external test to assess prospective teachers' knowledge of mathematics. Courses in effective programs include assessment strategies in alignment with approaches that prospective teachers will be expected to enact as future teachers (cf. NCTM, 2014a). This practice aligns with the increasing attention mathematicians are giving to the failure to connect what is valued and what is assessed (Tallman, Carlson, Bressoud, & Pearson, 2016). In an effective program, in addition to tests, other measures, such as projects and performance-based assignments, are used. Self-assessment is also a regular part of coursework, and formative assessment is an integral part of instruction and is used to make instructional decisions (AMTE & NCSM, 2014).

Beyond assessment within courses or equivalent professional-learning experiences, effective programs assess candidates' progress across courses. Quality assessment is more than an accumulation of individual course grades and includes assessment of dispositions, agencies, and understandings of the nature of mathematics. For example, tests of mathematical knowledge relevant to teaching—such as the Learning for Mathematics Teaching (2011) assessment or the Diagnostic Mathematics Assessments for Middle School Teachers (Peters et al., 2016a, 2016b; Ronau et al., 2016a, 2016b)—or assessments of dispositions toward mathematics might be incorporated at critical points in the program or as a requirement prior to student teaching. An effective program might also use a candidate-created work sample or electronic portfolio as an overall assessment of the depth and soundness of the candidate's knowledge of mathematics content and use of mathematical practices and processes.

Recommendation AC.2. Assessment of Mathematics Teaching Practice

Effective assessments of mathematics teaching practice include observations of teaching focused on how well the teaching supports learning of important mathematical content, processes, and practices by each and every student.

Effective mathematics teacher preparation programs offer candidates ongoing feedback from mentors and supervisors who know and have taught mathematics. These observers benefit from training and calibration using tools that are specific to the effective planning, teaching, and assessment practices highlighted in standard C.2, such as the Mathematics Classroom Observation Protocol for Practices (Gleason, Livers, & Zelkowski, 2015) and Student Discourse Observation Protocol (Weaver, Dick, & Rigelman, 2005). The observation cycle is most powerful when it includes preobservation discussion, focused observation of a lesson, data-based discussion following the lesson, and follow-up that holds candidates responsible for and supports their continued growth. The preobservation can engage the candidates in articulating their learning goals for students as well as potential foci for the observation (e.g., student engagement at various points in the lesson, responsiveness to student thinking/ideas, teacher questioning and wait time, student mathematical discourse, student use of mathematical tools). In the debrief of the lesson, participants use the data gathered by the observer to inform progress toward the learning goals and growth in instructional practice. Finally, following the observation cycle, the observer follows up with the candidate by checking on progress and providing resources to support continued development of teaching practice.

Recommendation AC.3. Assessment of Dispositions

Effective assessments provide data on a range of dispositions related to mathematics teaching, including dispositions toward engaging in mathematics, identity as a mathematics teacher and learner, and commitment to support the mathematics learning of each and every student.

Although candidates' reflections on their dispositions provide useful information and can serve as starting points for their continued growth, effective mathematics teacher preparation includes assessments of dispositions beyond candidates' self-assessment. Combining self-assessments with assessments from those working most closely with the candidates (i.e., advisers, faculty, mentor teachers, supervisors) can provide more powerful indications of their dispositions than self-assessments alone and can serve as points of reflection and growth for candidates. Dispositions can be further assessed when candidates reflect upon ways they support each of their students in developing productive mathematical identities.

ASSESSING QUALITY OF MATHEMATICS TEACHER PREPARATION PROGRAMS

In this section, we provide recommendations for how the effectiveness of mathematics teacher preparation programs in meeting the Program Standards from Chapter 3 might be assessed to guide efforts to improve, as outlined in Table 8.4. This table also shows that assessing the effectiveness of a program additionally requires attending to the recommendations for assessing candidate quality discussed in the previous section. In addition, the features of effective assessments outlined in the first section of this chapter guide decisions about the selection and use of assessments. The assessments used by effective programs are dominated by high-quality, valid assessments, results of which are used to spur and inform continuous program improvement.

Recommendation AP.1. Assessment of Stakeholder Engagement

Effective mathematics teacher preparation programs assess the degree to which a broad range of stakeholders are meaningfully engaged in the program and involved in its ongoing improvement.

Effective mathematics teacher preparation programs require the dedication and commitment of all those involved in preparing candidates. As described in Standard 3.1, although leadership for creating and sustaining the program may fall primarily to mathematics teacher educators, effective programs engage a range of other stakeholders, including colleagues in teacher preparation, Pre-K–12 educators and administrators, mathematicians, community-based organizations and community members, and state departments charged with the oversight of teacher preparation.

The diverse perspectives of all stakeholders are critical to program quality. Programs might use surveys, individual interviews, or focus groups to gather information from a range of stakeholders to answer questions such as the following:

- To what degree do the stakeholders believe that they are an important part of the mathematics teacher preparation program, that their perspectives are heard and acknowledged?
- To what degree do they perceive that the mathematics teacher preparation program is an important and meaningful part of their jobs and that their efforts contribute to the development of quality mathematics teacher candidates?
- To what degree do they feel an authentic partnership exists among the stakeholders, that they are not merely service providers to the program but are integral parts of the system?
- To what degree are the goals and values promulgated by the program shared across all those involved in the program?
- What areas of the program do they feel are functioning well? What areas need additional attention?

Stakeholders are further engaged in identifying and implementing assessments that provide information that they value and find useful in better understanding the effects of their policies and practices related to mathematics teacher preparation. The data collected are collaboratively considered to develop priorities for program improvement as well as to guide improvements in particular areas of the program.

Recommendation AP.2. Assessment of Program Curriculum and Instruction

Effective mathematics teacher preparation programs assess the effectiveness of the courses or equivalent professional-learning opportunities they offer in promoting their candidates' progress and work to improve them.

Although the typical course-evaluation procedures used by colleges and universities can provide limited information about the effectiveness of courses included in the mathematics teacher preparation program, effective mathematics teacher preparation programs undertake deeper assessments to ensure that all such courses or equivalent professional-learning experiences—including those addressing methods, mathematics content, and educational foundations—positively contribute to the growth of their candidates.

First, all courses or equivalent professional-learning experiences in effective mathematics teacher preparation programs model effective instructional approaches aligned with the instructional practices expected of well-prepared beginning teachers of mathematics (see Candidate Standards in Chapter 2). Instructors in key courses for mathematics teacher candidates self-assess their instruction, gather formative feedback from their candidates, and collaborate with others about how instruction can be improved. Peer observations using adaptations of observation protocols designed to observe the teaching of candidates might provide a useful starting point in such deliberations. Second, the content of courses in the program is analyzed to ensure that the courses promote candidate growth on the Candidate Standards. Course objectives and assignments are examined to ensure coherence across the program. Mapping the experiences provided onto the standards in this document, particularly P.2 and P.3, may help to identify areas in which more attention is needed or areas of overlap.

Finally, as stated in Standard P.3.5, effective programs assess whether instructors of mathematics methods experiences have relevant grade-level mathematics teaching experiences needed to support their candidates' growth. Although instructors of mathematics content courses need not have this same level of experience, the degree to which they are responsive to the content needs of mathematics teacher candidates is assessed, as outlined in Standard P.2. The awareness of instructors of other education foundational courses of how their content applies specifically to mathematics teaching and learning is also assessed.

Recommendation AP.3. Assessment of Effective Clinical Experiences

Effective mathematics teacher preparation programs assess and work to improve the quality of the clinical experiences they provide mathematics teacher candidates.

Effective assessment of the quality of the clinical experiences provided to mathematics teacher candidates requires consideration of the following, drawn from Standard P.4:

- Engagement of mentor teachers in enacting a shared vision of quality mathematics instruction. Effective programs need to assess what mechanisms are in place for ensuring bidirectional discussions with mentor teachers about mathematics teaching and learning. They need to engage mentor teachers in discussions about how clinical experiences can be better organized to drive continued improvement of the experiences offered.
- Effective sequencing of experiences. This sequencing might include examining requirements for experiences to ensure that they are coherently organized to respect the candidates' development and to become increasingly comprehensive in scope.

- Range of experiences. Records of the experiences candidates have should be analyzed to ensure experiences in a range of grade levels that reflect their certification levels as well as with students of varied backgrounds.
- Qualifications of mentors. Programs need to ensure that the Pre-K–12 mentor teachers and university supervisors reflect program values, demonstrate effective teaching, and have effective mentoring skills. University supervisors need to have relevant grade-level mathematics teaching experiences.

The assessments of effective programs also address the overall effectiveness of clinical experiences. This evidence might be gathered through specific, field-based assessments for individual candidates or across all candidates. For example, candidates' reflections on the effects of feedback on their professional growth provide insights into candidates' knowledge and skills, while also providing insights into the focus of the mentoring. If these collective data indicate little attention to particular areas (e.g., supporting and challenging each and every student), changes can be made to increase attention to those areas. Additionally, summative program data inform the quality of clinical experiences. These data might include reports of candidates' overall success in the program, surveys of both students and mentor teachers, and ratings of candidates on observation protocols. These analyses lead to discussions among university or program faculty and school personnel about areas in which clinical experiences support candidate growth and those in which adjustments may be needed.

Recommendation AP.4. Assessment of Recruitment and Retention

Effective mathematics teacher preparation programs assess and work to improve their success in recruiting prospective teachers, in retaining candidates in the program, and in supporting beginning teachers' continued professional growth, with a focus on developing a mathematics teacher workforce that reflects the diversity of the student population in the region.

The assessment systems of effective mathematics teacher preparation programs include expectations with measurable outcomes tied to effective recruitment and retention efforts, consistent with Standard 5 of the Council for the Accreditation of Educator Preparation (CAEP, 2013b). Effective programs assess the quality of their outcomes by setting goals for recruiting new mathematics teacher candidates and systematically gathering data and analyzing recruitment practices (e.g., effect on admissions from recruiting at high schools, community colleges, or other STEM majors.) Effective programs gather and review data on whether entrance requirements, which may include Praxis or other required candidate-assessment tests and GPA, are related to success in the program or whether other measures, such as flexibility in thinking and ability to relate to others unlike oneself, may, in fact, better predict a prospective teacher's potential to become a teacher of mathematics.

Effective programs assess factors that may influence program completion resulting in a well-prepared beginning teacher—for example, surveying candidates about the effectiveness of program supports while they progress through the program or conducting case studies of both successful and unsuccessful candidates. Effective programs monitor the early career progress of their graduates to improve programs as well as to contribute to the retention of teachers. Using social media might be one route through which programs maintain contact with graduates both to garner useful feedback and to create communities that provide peer and institutional support after program completion. Effective programs also assess practices for retaining beginning teachers in the profession, either those organized by the program or by the graduate's employer, to determine how those practices affect success of the beginning teacher. The program maintains data and examines results from other state or national teacher performance-assessment systems to ensure validity and examine how the data support or affect recruitment and retention as well as how results of the data analysis might inform program-improvement efforts.

Effective programs integrate attention to the diversity of their candidates in their assessments, with the goal of ensuring that the demographics of candidates completing their program are reflective of the student population of the region. For example, they might consider the following questions:

- Are current recruitment practices effective in reaching a diverse population of prospective teachers of mathematics?
- How might requirements for entry into the program be reconsidered to support increased diversity of mathematics teacher candidates accepted into the program?
- What support systems might be productive in ensuring that a diverse pool of candidates continue to make progress toward successful completion of the program and entry into the field as well-prepared beginning teachers of mathematics?
- What support systems could be useful in ensuring that program completers make progress as well-prepared beginning teachers of mathematics?

CLOSING REMARKS

The standards in Chapters 2 and 3 of this document set aggressive targets for the knowledge, skills, and dispositions of well-prepared beginning teachers of mathematics as well as for the qualities of effective mathematics teacher preparation programs. These standards are intended to guide efforts to support the development of well-prepared beginning teachers of mathematics and to improve mathematics teacher preparation programs. However, improvement is not possible without assessment. Evidence is necessary in identifying areas in which improvement is needed. Evidence also is needed to guide the process of improvement to ensure that changes are actually improvements (Bryk, Gomez, Grunow, & LaMahieu, 2015). Thus, seeking effective ways to assess the quality of candidates and of the program itself is central to attaining the goals set forth in this document. Indeed, when priorities for program improvement are considered, identifying or creating and implementing a more effective assessment system might be an important first step for a program to consider so that progress toward meeting the Candidate and Program Standards can be tracked.

CHAPTER 9. ENACTING EFFECTIVE PREPARATION OF TEACHERS OF MATHEMATICS

The *Standards for Preparing Teachers of Mathematics* has set forth an ambitious but achievable vision to prepare teachers who can effectively support the mathematics learning of each and every student in grades PK–12 education. Making the changes needed to achieve this vision will be challenging, requiring a significant investment of time and resources. Change, however, does not guarantee improvement, and improvement is imperative to meet the challenges facing mathematics teacher preparation outlined in the introduction to this document. In this final chapter, we offer suggestions to our AMTE members and the constituencies who partner with us for supporting the improvement of the preparation of teachers of mathematics, first reexamining the assumptions underlying improvement, then exploring processes that support improvement, and finally outlining necessary actions for stakeholders involved in preparing teachers of mathematics

IMPROVING THE PREPARATION OF TEACHERS OF MATHEMATICS

Efforts to improve mathematics teacher preparation must be rooted in the fundamental assumptions described in Chapter 1 and listed in Table 9.1.

TABLE 9.1. ASSUMPTIONS ABOUT PREPARING TEACHERS OF MATHEMATICS

Assumption	Statement
#1	Ensuring the success of each and every learner requires a deep, integrated focus on equity in every program that prepares teachers of mathematics.
#2	Teaching mathematics effectively requires career-long learning.
#3	Learning to teach mathematics requires a central focus on mathematics.
#4	Multiple stakeholders must be responsible for and invested in preparing teachers of mathematics.
#5	Those involved in mathematics teacher preparation must be committed to improving their effectiveness in preparing future teachers of mathematics.

Foremost among these assumptions is the attention to equity, reflecting a growing consensus within the mathematics education community of the need to confront an education system that marginalizes and disenfranchises a significant part of our citizenry (AMTE, 2015; NCTM, 2000, 2014a, 2014b). Throughout this document, attention to equity, diversity, and social justice issues has been a central focus, both specifically addressed as standards and embedded within all the standards. This focus has been maintained through the grade-band elaborations and in the discussion of assessment. Those involved in preparing teachers of mathematics cannot hope to make substantial improvement toward the vision of this document without placing issues of equity, diversity, and social justice front and center. Dealing with these issues with integrity requires more than a new slogan or policy statement. It requires honest and sometimes painful reflection on our beliefs, implicit assumptions and biases, and current practices in light of these standards to consider how our work affects the progress of the mathematics teacher candidates we are preparing in supporting the learning of each and every student they teach and how to innovate when necessary to improve what we are doing. Equity needs to become a lens through which all aspects of programs preparing teachers of mathematics are viewed. The question "How effective was this experience in cultivating the teacher candidate's ability to support each and every learner?" needs to become central when decisions about programs are made.

The other assumptions provide additional context for improvement efforts. Although the focus of this document has been on well-prepared beginning teachers, mathematics teacher preparation must be visualized as a continuum in which program completion is only the first step, as stated in the second assumption. A systemic view of teacher preparation is needed that extends from initial preparation to induction into the field to long-term professional learning. The third assumption states that although mathematics teacher preparation takes place within the context of teacher preparation more generally, the central focus on mathematics teaching and content must be maintained. Efforts successful for preparing teachers of reading, for example, may not be successful for preparing teachers of mathematics. The emphasis on partnerships found in the fourth assumption was reinforced in Standard P.1 and is also a focus of this final section. Efforts at improvement must engage not only mathematics teacher educators but also mathematicians; other teacher educators; program and school administrators; classroom teachers, including special education teachers; families and communities; policymakers; and others in the educational system.

Finally, this document should serve as a springboard for improvement, as stated in the final assumption. It should serve as a mirror in which all those involved in mathematics teacher preparation can closely examine what they are doing and how they can improve. Merely using the document to justify what is already happening while explaining away areas in which the vision is not being met will not be productive. Likewise, while focusing on what is effective can be encouraging, it will not ultimately spur the kind of deep reflection that leads to improvement. These standards should motivate a close look at what a program is accomplishing and how it can improve. Without a commitment to using the document to improve the practice of preparing teachers of mathematics, its purpose will not be realized.

PROCESS FOR SUPPORTING IMPROVEMENT OF MATHEMATICS TEACHER PREPARATION

In this section, we suggest practices or approaches that can be used to support the process of improving mathematics teacher preparation. Although making the changes needed to achieve the aspirational vision set forth in this document may be challenging, several frameworks have been developed to guide the implementation of a new set of recommendations. For example, Saul et al. (2008) presented the "Interactive Systems Framework for Dissemination and Implementation" to address the gap between research and practice in public health. Fixsen et al. (2005) described six levels of implementation based on a synthesis of research across multiple disciplines. Elrod and Kezar (2016) presented the "Model for Systemic Institutional Change" to support higher education leaders to guide change efforts in undergraduate STEM education. Taking a somewhat different approach, Bryk, Gomez, Grunow, and LeMahieu (2015) discussed the use of improvement science in networked communities to guide improvement in education, building on a body of work done in medicine.

These frameworks show that implementation is not a linear process but is more cyclic in nature while efforts are made to improve various aspects of the program. Figure 9.1 depicts the on-going and cyclic nature of successful improvement efforts, incorporating key elements found across the frameworks that should be considered while those involved with preparing teachers of mathematics work to implement the standards in this document.

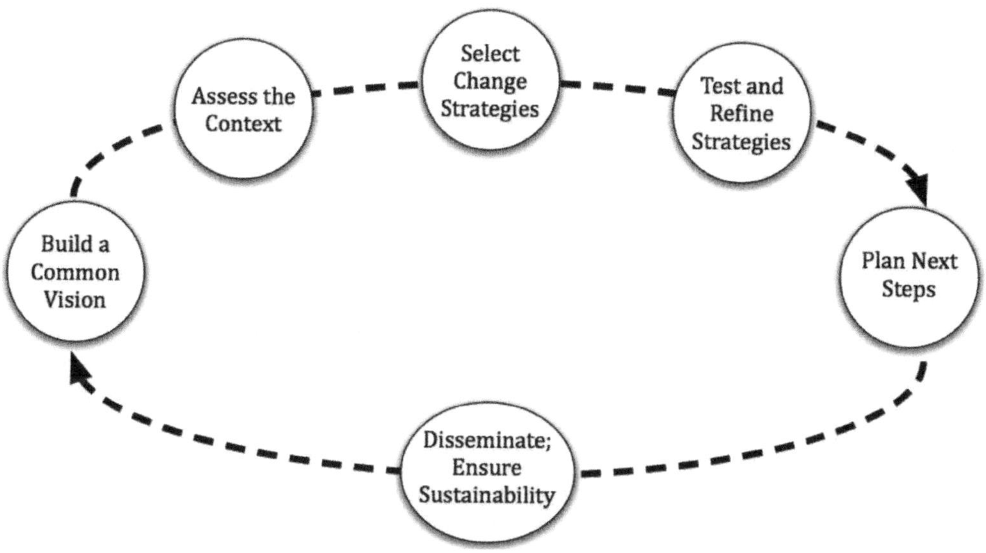

Figure 9.1. **The ongoing and cyclic nature of improving mathematics teacher preparation programs.**

Brief descriptions of the key elements follow:

- **Build a Common Vision**. Mathematics teacher preparation programs need to engage all stakeholders in establishing a common commitment to the standards, including universities, colleges, school districts, and schools, to devote the resources necessary to progress toward meeting the standards.
- **Assess the Context**. Efforts to improve the preparation of teachers of mathematics need to begin with a careful assessment of the context, examining current policies, structures, and outcomes in light of the standards to determine what practices are effective and what changes are needed.
- **Select Change Strategies**. As specific ideas for change are identified, consideration should be given to how ready team members are to make those changes before they are implemented. The process of identifying possible areas for change should be viewed as an ongoing process.
- **Test and Refine Strategies**. When changes are implemented, relevant data need to be collected to determine their efficacy and indicate any further adjustments that may be needed. These adjustments should then be tested, forming a cycle of innovation aimed at improving practice while allowing for the setbacks and delays that may sometimes occur.
- **Plan Next Steps.** Constant reflection on the improvement process is needed to plan next steps and maintain its momentum. Next steps could include selecting additional change strategies or revisiting any other elements in the cycle.
- **Disseminate.** Learning from and with other programs will accelerate the change process. Therefore, programs need to share successful innovations with the broader community.
- **Ensure Sustainability.** Programs need to ensure sustainability and institutionalization of the changes that are made so that the necessary resources continue to be devoted after an initial period of innovation. Furthermore, changes cannot be so dependent on particular innovators that they fade if those innovators leave.

Depending on the grain size of the changes being made, this cycle may stretch from one or more semesters to several years.

A CALL TO ACTION

Finally, improvement of mathematics teacher preparation programs must be seen within the context of the broad system in which the preparation of teachers of mathematics occurs; this system includes multiple constituencies, stakeholders, and organizations that influence teacher preparation. The challenges facing the preparation of teachers of mathematics are interconnected, and progress can be made only by addressing multiple audiences across the system. Several priority actions needed to begin to enact the vision of this document follow:

Action #1. Mathematics and statistics teacher educators must collaborate with mathematicians and statisticians to enact the vision for the content preparation of beginning teachers presented in these standards, building on related documents offered by other professional organizations (e.g., CBMS, 2012; Franklin et al., 2016; Tucker et al., 2015).

Action #2. Programs preparing beginning teachers of mathematics and Pre-K–12 schools and districts must develop close, respectful, bidirectional relationships that support the preparation of the next generation of teachers of mathematics, ensuring that school partners are included in the decision making about the preparation of beginning teachers.

Action #3. Mathematics educators and researchers must make these standards a focal point for research that will guide the improvement of mathematics teacher preparation. By identifying research questions and carrying out studies to support the development of mathematics teacher preparation programs, the research community will provide compelling ideas and evidence to ensure that program graduates are well-prepared beginning teachers of mathematics.

Action #4. Faculty in programs preparing teachers of mathematics must build collaborations with faculty in other programs preparing teachers of mathematics. Learning from and with colleagues from other institutions and providers can accelerate progress in their improvement efforts, with faculty benefitting from experiences and results of each site. The *networked improvement community* model proposed by Bryk, Gomez, Grunow, and LeMahieu (2015) may be particularly useful in building knowledge across programs (cf. Martin, W. G., & Gobstein, 2015).

Action #5. Higher education administrators overseeing programs that prepare teachers of mathematics must ensure that commitment and focus is placed on those programs, including allocating qualified personnel and resources needed to achieve the vision of this document. A program that does not make clear and strong commitments to ensuring that its candidates are well-prepared as described in this volume is not justifiable. Mediocrity cannot be an option.

Action #6. The Association of Mathematics Teacher Educators put forth this document to guide the improvement of mathematics teacher preparation. Accordingly this document must become a focal point for the continuing work of AMTE, including serving as an organizing theme of future conferences and publications. Committees and other working groups of the organization need to consider how their activities fit with the recommendations of this document.

Action #7. AMTE must work to engage other organizations with related missions in dialogue around how the standards can inform the preparation of teachers of mathematics. These organizations include other mathematics education organizations, mathematics teacher organizations with a primary focus on equity, mathematics organizations, organizations related to teaching learners of specific ages or other populations, and teacher education organizations. Finally, this document should serve as a primary resource for those involved in decisions regarding policies and accreditation standards for preparing teachers of mathematics.

CLOSING REMARKS

The *Standards for Preparing Teachers of Mathematics* has set forth an ambitious but achievable vision for the preparation of teachers of mathematics, grounded in fundamental assumptions that underlie the improvement of mathematics teacher preparation (see Table 9.1). Launching from these assumptions, these standards describe the knowledge, skills, and dispositions needed by well-prepared beginning teachers of mathematics (Chapter 2) and how to create effective mathematics teacher preparation programs to ensure that those successfully completing such programs are indeed well-prepared to meet the diverse needs of the students they will teach (Chapter 3). Elaborations of these standards describe specific guidance on preparing teacher candidates who will be teaching mathematics at different grade levels (Chapters 4 through 7), followed by recommendations for how to effectively assess progress in meeting these standards and elaborations (Chapter 8). Finally, in this chapter (Chapter 9), we make recommendations for enacting the vision.

We hope that these standards will evoke deep discussions among those involved in mathematics teacher preparation. We further hope that they will spur actions among those providing programs that prepare teachers of mathematics toward realizing the vision in their programs. Finally, we hope that these standards will shape the broader discourse around the need for continuous improvement in mathematics teacher preparation, serving as a seminal resource for reconsidering policies and practices that support mathematics teacher preparation.

REFERENCES

Aguirre, J., & Bunch, G. (2012). What's language got to do with it? Identifying language demands in mathematics instruction for English Language Learners. In S. Celedón-Pattichis & N. Ramirez (Eds.), *Beyond good teaching: Advancing mathematics education for ELLs* (pp. 183–194). Reston, VA: National Council of Teachers of Mathematics.

Aguirre, J. M., Mayfield-Ingram, K., & Martin, D. (2013). *The impact of identity in K–8 mathematics learning and teaching: Rethinking equity-based practices.* Reston, VA: National Council of Teachers of Mathematics.

Aguirre, J. M., Turner, E. E., Bartell, T., Kalinec-Craig, C., Foote, M. Q., Roth McDuffie, A., & Drake, C. (2012). Making connections in practice: How prospective elementary teachers connect children's mathematics thinking and community funds of knowledge in mathematics instruction. *Journal of Teacher Education, 64*(2), 178–192.

Aguirre, J. M., & Zavala, M. (2013). Making culturally responsive mathematics teaching explicit: A lesson analysis tool. *Pedagogies: An International Journal*, *8*(2), 163–190.

Ahmad, F. Z., & Boser, U. (2014). *America's leaky pipeline for teachers of color: Getting more teachers of color into the classroom.* Washington, DC: Center for American Progress.

American Statistical Association. (2007). *Guidelines for assessment and instruction in statistics education: A pre-K–12 curriculum framework.* Retrieved from http://www.amstat.org/asa/files/pdfs/GAISE/GAISEPreK-12_Full.pdf

American Statistical Association and National Council of Teachers of Mathematics. (2015). *Preparing pre-K–12 teachers of statistics. A joint statement of ASA and NCTM.* Retrieved from http://www.nctm.org/Standards-and-Positions/Position-Statements/Preparing-Pre-K-12-Teachers-of-Statistics/

Association for Middle Level Education (AMLE). (2010). *This we believe: Keys to educating young adolescents*. Westerville, OH: Author.

Association for Middle Level Education (AMLE). (2012). *Association for middle level education: Middle level teacher preparation standards with rubrics and supporting explanations.* Retrieved from http://www.amle.org/AboutAMLE/ProfessionalPreparation/AMLEStandards.aspx

Association for Middle Level Education (AMLE). (2015). *Middle level teacher preparation and certification/licensure.* Retrieved from https://www.amle.org/BrowsebyTopic/WhatsNew/WNDet/TabId/270/ArtMID/888/ArticleID/487/Middle-Level-Teacher-Preparation-and-CertificationLicensure.aspx

Association of Mathematics Teacher Educators. (2013). *Standards for elementary mathematics specialists: A guide for teacher credentialing and degree programs.* Retrieved from http://amte.net/sites/all/themes/amte/resources/EMS_Standards_AMTE2013.pdf

Association of Mathematics Teacher Educators. (2015). *Position: Equity in mathematics teacher preparation*. Retrieved from https://amte.net/sites/default/files/amte_equityposistionstatement_sept2015.pdf

Association of Mathematics Teacher Educators and National Council of Teachers of Mathematics. (2014). *Improving student achievement in mathematics through formative assessment in instruction.* Raleigh, NC: Author. Retrieved from https://www.amte.net/resources/formative-assessment

Bacharach, N., Heck, T. W., & Dahlberg, K. (2010). Changing the face of student teaching through co-teaching. *Action in Teacher Education*, *32*(1), 3–14.

Ball, D. L. (1992). Magical hopes: Manipulatives and the reform of mathematics education. *American Educator, 16*(2), 14–18, 46–47.

Ball, D. L., & Cohen, D. K. (1999). Developing practice, developing practitioners: Toward a practice-based theory of professional education. In L. Darling-Hammond & L. Sykes (Eds.), *Teaching as the learning profession: Handbook of policy and practice* (pp. 3–32). San Francisco, CA: Jossey-Bass.

Ball, D. L., & Forzani, F. M. (2009). The work of teaching and the challenge for teacher education. *Journal of teacher education*, *60*(5), 497–511.

Ball, D. L., & Forzani, F. M. (2011). Building a common core for learning to teach, and connecting professional learning to practice. *American Educator, 35*(2), 17–21, 38–39.

Ball, D. L., Thames, M. H., & Phelps, G. (2008). Content knowledge for teaching: What makes it special? *Journal of Teacher Education, 59*(5), 389–407.

Barnett-Clarke, C., Fisher, W., Marks, R., & Ross, S. (2010). Developing essential understanding of rational numbers for teaching mathematics in Grades 3–5 (R. I. Charles, Ed.). In R. M. Zbiek (Series Ed.), *Essential understanding series.* Reston, VA: National Council of Teachers of Mathematics.

Baron, R., Tom, D., & Cooper, H. M. (1985). Social class, race and teacher expectations. In J. B. Dusek (Ed.), *Teacher Expectancies*. Hillsdale, NJ: Erlbaum.

Baroody, A. J. (1999). Children's relational knowledge of addition and subtraction. *Cognition and Instruction, 17*(2), 137–175.

Bay-Williams, J., McGatha, M., Kobett, B., & Wray, J. (2014). *Mathematics coaching: Resources and tools for coaches and leaders, K–12.* New York, NY: Pearson.

Bell, C. A., Horn, B. R, & Roxas, K. C. (2007). We know it's service, but what are they learning? Preservice teachers' understanding of diversity. *Equity and Excellence in Education, 40*(2), 123–133.

Berry, R. Q., III. (2008). Access to upper-level mathematics: The stories of African American middle school boys who are successful with school mathematics. *Journal for Research in Mathematics Education, 39*(5), 464–488.

Berry, R. Q., III, Ellis, M., & Hughes, S. (2014). Examining a history of failed reforms and recent stories of success: Mathematics education and Black learners of mathematics in the United States. *Race, Ethnicity, and Education, 17*(4), 540–568.

Black, P., & Wiliam, D. (1998). Inside the black box: Raising standards through classroom assessment. *Phi Delta Kappan, 80*(2), 139–148.

Blanton, M., Levi, L., Crites, T., & Dougherty, B. J. (2011). Developing essential understanding of algebraic thinking for teaching mathematics in Grades 3–5 (B. J. Dougherty, Ed.). In R. M. Zbiek (Series Ed.), *Essential understanding series.* Reston, VA: National Council of Teachers of Mathematics.

Boaler, J. (2011). Changing students' lives through the de-tracking of urban mathematics classrooms. *Journal of Urban Mathematics Education, 4*(1), 7–14.

Boaler, J. (2016). *Mathematical mindsets: Unleashing students' potential through creative math, inspiring messages, and innovative teaching.* San Francisco, CA: Jossey-Bass.

Bol, L., & Berry, R. Q., III. (2005). Secondary mathematics teachers' perceptions of the achievement gap. *High School Journal, 88*(4), 32–45.

Boyd, D., Grossman, P., Lankford, H., Loeb, S., & Wyckoff, J. (2009). Teacher preparation and student achievement. *Educational Evaluation and Policy Analysis, 31*(4), 416–440.

Brenneman, K., Stevenson-Boyd, J., & Frede, E. C. (2009). Math and science in preschool: Policies and practice. *Preschool Policy Brief, 19*, 1–12.

Bressoud, D., Mesa V., & Rasmussen, C. (2015). *Insights and recommendations from the MAA National Study of College Calculus*. Retrieved from http://www.maa.org/sites/default/files/pdf/cspcc/InsightsandRecommendations.pdf

Bryk, A., Gomez, L. M., Grunow, A., & LeMahieu, P. (2015). *Learning to improve: How America's schools can get better at getting better.* Cambridge, MA: Harvard Education Press.

Cai, J. (2003). What research tells us about teaching mathematics through problem solving. In F. Lester (Ed.), *Research and issues in teaching mathematics through problem solving* (pp. 241–254). Reston, VA: National Council of Teachers of Mathematics.

Caldwell, J. H., Karp, K., & Bay-Williams, J. M. (2011). Developing essential understanding of addition and subtraction for teaching mathematics in Pre-K–Grade 2 (E. Rathmell, Ed.). In R. M. Zbiek (Series Ed.), *Essential understanding series.* Reston, VA: National Council of Teachers of Mathematics.

Carpenter, T. P., Fennema, E., Franke, M. L., Levi, L., & Empson, S. B. (2014). *Children's mathematics* (2nd ed.), Portsmouth, NH: Heinemann.

Carpenter, T. P., Fennema, E. H., Peterson, P. L., & Carey, D. A. (1988). Teacher's pedagogical content knowledge of students' problem solving in elementary arithmetic. *Journal for Research in Mathematics Education, 19*, 385–401.

Carpenter, T. P., Franke, M. L., & Levi, L. (2003). *Thinking mathematically: Integrating arithmetic & algebra in elementary school.* London, England: Heinemann.

Casa, T. M., Firmender, J. M., Cahill, J., Cardetti, F., Choppin, J. M., Cohen, J., ... Zawodniak, R. (2016). *Types of and purposes for elementary mathematical writing: Task force recommendations.* Retrieved from http://mathwriting.education.uconn.edu.

Castro, A. J. (2010). Themes in the research on preservice teachers' views of cultural diversity: Implications for researching millennial preservice teachers. *Educational Researcher*, *39*(3), 198–210.

Celedón-Pattichis, S., & Ramirez, N. G. (2012). *Beyond good teaching: Advancing mathematics education for ELLs*. Reston, VA: National Council of Teachers of Mathematics.

Chance, B. (2002). Components of statistical thinking and implications for instruction and assessment. *Journal of Statistics Education, 10* (3). Retrieved from https://ww2.amstat.org/publications/JSE/v10n3/chance.html

Charles, R. I. (2005). Big ideas and understandings as the foundation for elementary and middle school mathematics. *Journal of Mathematics Education Leadership*, *7*(3), 9–24.

Chval, K. B., Lannin, J., Arbaugh, F., & Bowzer, A. (2009). Videos and prospective teachers. *Teaching Children Mathematics, 16*(2), 98–105.

Chval, K. B., & Pinnow, R. (2010). Preservice teachers' assumptions about Latino/a English language learners. *Journal of Teaching for Excellence and Equity in Mathematics*, *2*(1), 6–12.

Civil, M. (2007). Building on community knowledge: An avenue to equity in mathematics education. In N. Nasir & P. Cobb (Eds.), *Improving access to mathematics: Diversity and equity in the classroom* (pp. 105–117). New York, NY: Teachers College Press.

Civil, M. (2016). "This is nice but they need to learn to do things the U.S. way": Reactions to different algorithms. In D. Y. White, S. Crespo, & M. Civil (Eds.), *Cases of teacher educators: Facilitating conversations about inequities in mathematics classrooms* (pp. 219–226). Charlotte, NC: Information Age.

Civil, M., & Bernier, E. (2006). Exploring images of parental participation in mathematics education: Challenges and possibilities. *Mathematical Thinking and Learning*, *8*(3), 309–330.

Civil, M., & Menéndez, J. M. (2011). Impressions of Mexican immigrant families on their early experiences with school mathematics in Arizona. In R. Kitchen & M. Civil (Eds.), *Transnational and borderland studies in mathematics education* (pp. 47–68). New York, NY: Routledge.

Civil, M., & Planas, N. (2010). Latino/a immigrant parents' voices in mathematics education. In E. Grigorenko & R. Takanishi (Eds.), *Immigration, diversity, and education* (pp. 130–150). New York, NY: Routledge.

Clarke, C., Fisher, W., Marks, R., Ross, S., Zbiek, R. M. (2010). *Developing essential understanding of rational numbers for teaching mathematics in grades 3–5.* Reston, VA: National Council of Teachers of Mathematics.

Clarke, D., Clarke, B., & Roche, A. (2011). Building teachers' expertise in understanding, assessing and developing children's mathematical thinking: The power of task-based, one-to-one interviews. *ZDM Mathematics Education, 43*(6), 901–913.

Clements, D. H., & Sarama, J. (2014). *Learning and teaching early math: The learning trajectories approach* (2nd ed.). New York, NY: Routledge.

Clift, R., & Brady, P. (2006). Research on methods courses and field experiences. In M. Cochran-Smith & K. Zeichner (Eds.), *Studying teacher education: The report on the AERA panel on Research and Teacher Education* (pp. 309–424). Mahwah, NJ: Erlbaum.

Coble, C. R. (2012). *Developing the analytic framework: A tool for supporting innovation and quality design in the preparation and development of science and mathematics teachers.* Washington, DC: Association of Public and Land-grant Universities.

Cogan, L. S., Schmidt, W. H., & Wiley, D. E. (2001). Who takes what math and in which track? Using TIMSS to characterize U.S. students' eighth-grade mathematics learning opportunities. *Educational Evaluation and Policy Analysis*, *23*(4), 323–341.

Conaway, B. J., Browning, L. J., & Purdum-Cassidy, B. (2007). Teacher candidates' changing perceptions of urban schools: Results of a 4-year study. *Action in Teacher Education, 29*(1), 20–31.

Conference Board of the Mathematical Sciences (CBMS). (2012). *The mathematical education of teachers II*. Providence, RI, and Washington, DC: American Mathematical Society and Mathematical Association of America.

Consortium for Mathematics and Its Applications & Society for Industrial and Applied Mathematics (CMA & SIAM). (2016). *Guidelines for assessment and instruction in mathematical modeling Education*. Retrieved from http://www.siam.org/reports/gaimme.php

Council for Exceptional Children (CEC). (2012). *CEC Initial Level Special Educator Preparation Standards*. Retrieved from http://www.cec.sped.org/~/media/Files/Standards/Professional%20Preparation%20Standards/Initial%20Preparation%20Standards%20with%20Elaborations.pdf

Council for the Accreditation of Educator Preparation (CAEP) (2013). *CAEP Accreditation Standards*. Retrieved from http://caepnet.org/~/media/Files/caep/standards/caep-2013-accreditation-standards.pdf

Council of Chief State School Officers (CCSSO). (2013, April). Interstate Teacher Assessment and Support Consortium InTASC *Model Core Teaching Standards and Learning Progressions for Teachers 1.0: A resource for ongoing teacher development*. Washington, DC: Author.

Darling-Hammond, L. (2000). How teacher education matters. *Journal of Teacher Education 51*(3), 166–173.

Darling-Hammond, L., & Bransford, J. (2007). *Preparing teachers for a changing world: What teachers should learn and be able to do*. Hoboken, NJ: John Wiley & Sons.

Daro, P., Mosher, F. A., & Corcoran, T. (2011). *Learning trajectories in mathematics: A foundation for standards, curriculum, assessment, and instruction*. Philadelphia, PA: Consortium for Policy Research in Education.

Davis, E. A., & Krajcik, J. S. (2005). Designing educative curriculum materials to promote teacher learning. *Educational Researcher, 34*(3), 3–14.

De Carvalho Borba, M., Villareal, M. E., & da Silva Soares, D. (2016). Modeling using data available on the Internet. In C. R. Hirsch & A. Roth McDuffie (Eds.), *Annual perspectives in mathematics education: Mathematical modeling and modeling mathematics* (pp. 143–152). Reston, VA: NCTM.

Dewey, J. (1910). Science as subject-matter and as method. *Science, 31*, 121–127.

Dick, T., & Hollebrands, K. (2011). *Focus in high school mathematics: Technology to support reasoning and sense making.* Reston, VA: National Council of Teachers of Mathematics.

Dickey, E. (2016, January). *Rebranding the teaching profession: Ideas and strategies for effective recruitment of mathematics teachers.* Opening of general session of Association of Mathematics Teacher Education 20th annual meeting, Irvine, CA.

Doerr, H., Goldsmith, L., & Lewis, C. (2010). *Mathematics professional development* [Research Brief]. Reston, VA: National Council of Teachers of Mathematics.

Doig, B., McCrae, B., & Rowe, K. (2003). *A good start to numeracy: Effective numeracy strategies from research and practice in early childhood.* Retrieved from http://research.acer.edu.au/learning_processes/3/

Domínguez, H. (2011). Using what matters to students in bilingual mathematics problems. *Educational Studies in Mathematics,* 76(3), 305–328.

Dougherty, B. J., Flores, A., Louis, E., & Sophian, C. (2010). *Developing essential understanding of number and numeration, Pre-K–Grade 2.* R. M. Zbiek (Series Ed.), Essential understanding series. Reston, VA: National Council of Teachers of Mathematics.

Duncan, G. J., Dowsett, C. J., Claessens, A., Magnuson, K., Huston, A. C., Klebanov, P., & Japel, C. (2007). School readiness and later achievement. *Developmental Psychology, 43*(6), 1428–1446. doi: 10.1037/0012-1649.43.6.1428

Dweck, C. S. (2008). *Mindsets and math/science achievement*. New York: Carnegie Corporation of New York Institute for Advanced Study.

Elrod, S., & Kezar, A. (2016). *Increasing student success in STEM: A guide to systemic institutional change.* Washington, DC: Association of American Colleges & Universities.

Ercole, L. K., Frantz, M., & Ashline, G. (2011). Multiple ways to solve proportions. *Mathematics Teaching in the Middle School, 16*(8), 482–490.

Esmonde, I., & Langer-Osuna, J. (2013). Power in numbers: Student participation in mathematical discussions in heterogeneous spaces. *Journal for Research in Mathematics Education, 44*(1), 288–315.

Falkner, K., Levi, L., & Carpenter, T. (1999). Children's understanding of equality: A foundation for algebra. *Teaching Children Mathematics, 6*(4), 232–236.

Feuer, M. J., Floden, R. E., Chudowsky, N., & Ahn, J. (2013). *Evaluation of teacher preparation programs: Purposes, methods, and policy options*. Washington, DC: National Academy of Education.

Fixsen, D. L., Naoom, S. F., Blase, K. A., Friedman, R. M., & Wallace, F. (2005). Implementation research: A synthesis of the literature. Tampa, FL: University of South Florida, Louis de la Parte Florida Mental Health Institute, NIRN (FMHI Publication #231).

Foote, M. Q. (2009). Stepping out of the classroom: Building teacher knowledge for developing classroom practice. *Teacher Education Quarterly, 36*(3), 39–53.

Forzani, F. M. (2014). Understanding "core practices" and "practice-based" teacher education: Learning from the past. *Journal of Teacher Education, 65*(4), 357–368.

Franklin, C., Bargagliotti, A., Case, C., Kader, G., Schaeffer, R., & Spangler, D. (2015). *Statistical education of teachers*. Arlington, VA: American Statistical Association. Retrieved from http://www.amstat.org/

Freeman, S., Eddy, S., McDonough, M., Smith, M., Okoroafor, N., Jordt, H., & Wenderoth, M. P. (2014). Active learning increases student performance in science, engineering, and mathematics. *Proceedings of the National Academy of Sciences, 111*(23) 8410–8415.

Fulton, K., Doerr, H., & Britton, T. (2010). *STEM teachers in professional learning communities: A knowledge synthesis*. Washington, DC: National Commission on Teaching & America's Future & West Ed.

Fuson, K. C., Kalchman, M., & Bransford, J. D. (2005). Mathematical understanding: An introduction. In M. S. Donovan & J. D. Bransford (Eds.), *How students learn: History, mathematics, and science in the classroom* (pp. 217–256). Washington, DC: National Academies Press.

Garmon, M. A. (2004). Changing preservice teachers' attitudes and beliefs about diversity: What are the critical factors? *Journal of Teacher Education, 55*(3), 201–213.

Generating Increased Science and Mathematics Opportunities. (2012). *Learning trajectories for the K–8 Common Core Math Standards.* Retrieved from https://www.turnonccmath.net/

Gleason, J., Livers, S. D., & Zelkowski, J. (2015). *Mathematics classroom observation protocol for practices: Descriptors manual*. Retrieved from http://jgleason.people.ua.edu/mcop2.html

Gojak, L. (2013). Algebra: Not 'if' but "when." *NCTM summing up*. Retrieved from http://www.nctm.org/News-and-Calendar/Messages-from-the-President/Archive/Linda-M_-Gojak/Algebra_-Not-_If_-but-_When_/

Goldhaber, D., & Hansen, M. (2010). Race, gender, and teacher testing: How informative a tool is teacher licensure testing? *American Educational Research Journal, 47*(1), 218–251.

González, N., Andrade, R., Civil, M., & Moll, L. (2001). Bridging funds of distributed knowledge: Creating zones of practices in mathematics. *Journal of Education for Students Placed at Risk, 6*(1–2), 115–132.

Goodnough, K., Osmond, P., Dibbon, D., Glassman, M., & Stevens, K. (2009). Exploring a triad model of student teaching: Preservice teacher and cooperating teacher perceptions. *Teaching and Teacher Education, 25*(2), 285–296.

Gravemeijer, K. (2004). Local instruction theories as means of support for teachers in reform mathematics education. *Mathematical thinking and learning, 6*(2), 105–128.

Grossman, P. (1990). *The making of a teacher: Teacher knowledge and teacher education.* New York, NY: Teachers College Press.

Grossman, P. (2010, May). *Learning to practice: The design of clinical experience*. Washington, DC: American Association of Colleges for Teacher Education & National Education Association.

Grossman, P., Compton, C., Igra, D., Ronfeldt, M., Shahan, E., & Williamson, P. W. (2009). Teaching practice: A cross-professional perspective. *Teachers College Record, 111*(9), 2055–2100.

Grossman, P., Ronfeldt, M., & Cohen, J. (2011). The power of setting: The role of field experience in learning to teach. In K. Harris, S. Graham, T. Urdan, A. Bus, S. Major, & H. L. Swanson (Eds.), *American Psychological*

Association (APA) Educational Psychology Handbook: Vol. 3. Applications to teaching and learning (pp. 311–334). Washington DC: American Psychological Association.

Guillaume, A. M., & Kirtman, L. (2005). Learning lessons about lessons: Memories of mathematics instruction. *Teaching Children Mathematics, 11*(6), 302.

Gutiérrez, R. (2009). Embracing the inherent tensions in teaching mathematics from an equity stance. *Democracy and Education, 18*(3), 9–16.

Gutiérrez, R. (2013a). The sociopolitical turn in mathematics education. *Journal for Research in Mathematics Education*, *44*(1), 37–68.

Gutiérrez, R. (2013b). Why (urban) mathematics teachers need political knowledge. *Journal of Urban Mathematics Education*, *6*(2) 7–19.

Gutiérrez, R. (2015). HOLA: Listening to Latin@ students. *Mathematics Teacher*, *109*(4), 271–277.

Gutiérrez, R. (2016). Strategies for creative insubordination in mathematics teaching. *Teaching for Excellence and Equity in Mathematics*, 7(1), 52–60.

Gutiérrez, R., Bay-Williams, J., & Kanold, T. D. (2008). Moving beyond access and achievement: What should mathematics teachers and leaders consider when addressing equity issues? *NCTM News Bulletin, 1,* 3. Reston: National Council of Teachers of Mathematics.

Gutiérrez, R., Gerardo, J. M., & Vargas, G. V. (in press). Rehearsing for the politics of teaching mathematics. To appear in S. Kastberg, A. M. Tyminski, A. Lischka, & W. Sanchez (Eds.), *Building support for scholarly practices in mathematics methods*. Raleigh, NC: Association of Mathematics Teacher Educators.

Gutiérrez, R., & Gregson, S. A. (2013, April). *Mathematics teachers and creative insubordination: Taking a stand in high-poverty schools.* Selected paper presented at the annual meeting of the American Educational Research Association. San Francisco, CA.

Gutstein, R. (2006). *Reading and writing the world with mathematics: Toward a pedagogy for social justice*. New York, NY: Routledge.

Hendrickson, S., Siebert, D., Smith, S. Z., Kunzler, H. & Christensen, S. (2004). Addressing parents' concerns about mathematics reform. *Teaching Children Mathematics, 11*(1), 18–23.

Henry, V. J., & Brown, R. S. (2008). First-grade basic facts: An investigation into teaching and learning of an accelerated, high-demand memorization standard. *Journal for Research in Mathematics Education*, *39*(2), 153–183.

Hiebert, J., Morris, A. K., Berk, D., & Jansen, A. (2007). Preparing teachers to learn from teaching. *Journal of Teacher Education, 58*(1), 47–61.

Hiebert, J., Stigler, J. W., Jacobs, J. K., Givvin, K. B., Garnier, H., Smith, M., Hollingsworth, H., Manister, A., Wearne, D., & Gallimore, R. (2005). Mathematics teaching in the United States today (and tomorrow): Results from the TIMSS 1999 video study. *Educational Evaluation and Policy Analysis, 27*(2), 111–132.

Hiebert, J., & Wearne, D. (1993). Instructional tasks, classroom discourse, and students' learning in second-grade arithmetic. *American Educational Research Journal 30*(2), 393–425.

Hill, H. C., Rowan, B., & Ball, D. (2005). Effects of teachers' mathematical knowledge for teaching on student achievement. *American Educational Research Journal, 42* (2), 371–406.

Hirsch, C., & Roth McDuffie, A. (Eds.). (2016). *Annual perspectives in mathematics education 2016: Mathematical modeling and modeling mathematics.* Reston, VA: National Council of Teachers of Mathematics.

Hodges, T. E., Rose, T. D., & Hicks, A. D. (2012). Interviews as RtI tools. *Teaching Children Mathematics, 19*(1), 30–36.

Holm, J., & Kajander, A. (2012). "I finally get it!": Developing mathematical understanding during teacher education. *International Journal of Mathematics Education in Science and Technology*, *43*(5), 563–574.

Hong, L.T. (1993). *Two of everything.* Morton Grove, IL: Albert Whitman.

Horn, I. S. (2006). Lessons learned from detracked mathematics departments. *Theory Into Practice*, *45*(1), 72–81.

Horn, I. S. (2010). Teaching replays, teaching rehearsals, and revisions of practice: Learning from colleagues in a mathematics teacher community. *Teachers College Record, 112*(1), 225–259.

Huinker, D. (2011). Beyond counting by ones: "Thinking in groups" as a foundation for number and operation sense. *Wisconsin Teacher of Mathematics, 63(1)*, 7–11.

Individuals With Disabilities Education Act. (n.d.). Retrieved from http://idea.ed.gov/explore.

Institute for Mathematics and Education. (n.d.). *Progressions documents for the Common Core math standards.* Retrieved from http://math.arizona.edu/~ime/progressions/

Institute of Medicine (IOM) and National Research Council [NRC]. (2015). *Transforming the workforce for children birth through age 8: A unifying foundation*. Washington, DC: National Academy Press.

Isenberg, J. P. (2000). The state of the art in early childhood professional preparation. In D. Horm-Wingerd & M. Hyson (Eds.), *New teachers for a new century: The future of early childhood professional preparation* (pp. 17–58). Washington, DC: U.S. Department of Education.

Jackson, K., Garrison, A., Wilson, J., Gibbons, L., & Shahan, E. (2013). Exploring relationships between setting up complex tasks and opportunities to learn in concluding whole-class discussions in middle grades mathematics instruction. *Journal for Research in Mathematics Education, 44*(4), 646–682.

Jacobs, V. R., Lamb, L. L. C., & Philipp, R. A. (2010). Professional noticing of children's mathematical thinking. *Journal for Research in Mathematics Education, 41*(2), 169–202.

Jong, C., & Hodges, T. E. (2015). Assessing attitudes toward mathematics across teacher education contexts. *Journal of Mathematics Teacher Education, 18*(5), 407–425.

Kahan, D. M., Peters, E., Dawson, E. C., & Slovic, P. (2013). *Motivated numeracy and enlightened self-government.* The Cultural Cognition Project. Working Paper 116. Retrieved from http://papers.ssrn.com/sol3/papers.cfm?abstract_id=2319992

Kalogrides, D., Loeb, S., & Béteille, T. (2012). Systematic sorting: Teacher characteristics and class assignments. *Sociology of Education, 86*(2), 103–123.

Kaput, J. J. (2008). What is algebra? What is algebraic reasoning? In J. J. Kaput, D. W. Carraher, & M. L. Blanton (Eds.), *Algebra in the early grades* (pp. 5–17). Reston, VA: National Council of Teachers of Mathematics.

Karp, K. S., Bush, S. B., & Dougherty, B. J. (2014, August). 13 rules that expire. *Teaching Children Mathematics, 21*(1), 18–25.

Kellough, R. D., & Kellough, N. G. (2008). *Teaching young adolescents: Methods and resources for middle grades teaching* (5th ed.). Upper Saddle River, NJ: Pearson Merrill Prentice Hall.

Kennedy, M. (1998). *Form and substance of in-service teacher education (Research Monograph No. 13).* Madison, WI: National Institute for Science Education, University of Wisconsin–Madison.

Kilpatrick, J., Swafford, J., & Findell, B. (2001). *Adding it up: Helping children learn mathematics.* Washington, DC: National Academy Press.

Kwong, C. W., Joseph, Y. K. K., Eric, C. C. M., Khoh, L.-T. S., Gek, C. K., & Eng, N. L. (2007). Development of mathematics pedagogical content knowledge in student teachers. *The Mathematics Educator, 10*(2), 27–54.

Lamon, S. J. (2012). Teaching fractions and ratios for understanding: *Essential content knowledge and instructional strategies* (3rd ed.). New York, NY: Taylor & Francis.

Lampert, M. (2001). *Teaching problems and the problems of teaching*. New Haven, CT: Yale University Press.

Lampert, M., Boerst, T., & Graziani, F. (2011). Organizational resources in the service of school-wide ambitious teaching practice. *Teachers College Record, 113*(7), 1–20.

Lampert, M., Franke, M., Kazemi, E., Ghousseini, H., Turrou, A., Beasley, H., Cunard, A., & Crowe, K. (2013). Keeping it complex : Using rehearsals to support novice teacher learning of ambitious teaching. *Journal of Teacher Education, 64*(3), 226–243.

Lampert, M., & Graziani, F. (2009). Instructional activities as a tool for teachers' and teacher educators' learning. *The Elementary School Journal, 109*(5), 491–509.

Larnell, G. V. (2016). More than just skill: Examining mathematics identities, racialized narratives, and remediation among black undergraduates. *Journal for Research in Mathematics Education, 47*(3), 233–269.

Learning Mathematics for Teaching. (2011). Measuring the mathematical quality of mathematics teaching. *Journal for Mathematics Teacher Education 14*(1), 25–47.

Leatham, K. R., & Peterson, B. E. (2010). Purposefully designing student teaching to focus on students' mathematical thinking. In J. W. Lott & J. Luebeck (Eds.), *Mathematics teaching: Putting research into practice at all levels* (pp. 225–239). San Diego, CA: Association of Mathematics Teacher Education.

Lee, J. S., & Ginsburg, H. P. (2009). Early childhood teachers' misconceptions about mathematics education for young children in the United States. *Australasian Journal of Early Childhood, 34*(4), 37–46.

Lehrer, R., & Slovin, H. (2014). Developing essential understanding of geometry and measurement for teaching mathematics in Grades 3–5 (B. J. Dougherty, Ed.). In R. M. Zbiek (Series Ed.), *Essential understanding series.* Reston, VA: National Council of Teachers of Mathematics.

Leonard, J., Brooks, W., Barnes-Johnson, J., & Berry, R. Q., III. (2010). The nuances and complexities of teaching mathematics for cultural relevance and social justice. *Journal of Teacher Education, 61*(3), 261–270.

Leonard, J. & Martin, D. B. (2013). The brilliance of Black children in mathematics: Beyond the numbers and towards a new discourse. Charlotte, NC: Information Age.

Lerkkanen, M. K., Rasku-Puttonen, H., Aunola, K., & Nurmi, J. E. (2005). Mathematical performance predicts progress in reading comprehension among 7-year-olds. *European Journal of Psychology of Education, 20*(2), 121–137.

Lloyd, G., & Beckmann-Kazez, S. (2010). Developing essential understanding of functions for teaching mathematics in Grades 9–12. In R. M. Zbiek (Series Ed.), *Essential understanding series.* Reston, VA: National Council of Teachers of Mathematics.

Lloyd, G., Herbel-Eisenmann, B., & Star, J. R. (2011). Developing essential understanding of expressions, equations, and functions for teaching mathematics in Grades 6–8. In R. M. Zbiek (Series Ed.), *Essential understanding series.* Reston, VA: National Council of Teachers of Mathematics.

Loveless, T. (2013). *The 2013 Brown Center report on American education: How well are American students learning?* Houston, TX: Brown Foundation.

Loveless, T. (2016). *How well are American students learning? Part II: Tracking and advanced placement* (Vol., 3 (5), pp. 16–25). Retrieved from http://www.brookings.edu/~/media/Research/Files/Reports/2016/03/brown-center-report/Brown-Center-Report-2016.pdf?la=en

Lovett, J. N., & Lee, H. S. (2016). Making sense of data: Context matters. *Mathematics Teaching in the Middle School, 21*(6), 338–346.

Ma, L. (1999). *Knowing and teaching elementary mathematics: Teachers' understanding of fundamental mathematics in China and the United States*. Mahwah, NJ: Erlbaum.

Martin, D. B. (2015). The collective Black and principles to actions. *Journal of Urban Mathematics Education, 8*(1), 17–23.

Martin, D. B., Gholson, M. L., & Leonard, J. (2010). Mathematics as gatekeeper: Power and privilege in the production of knowledge. *Journal of Urban Mathematics Education, 3*(2), 12–24.

Martin, W. G., & Gobstein, H. (2015). Generating a networked improvement community to improve secondary mathematics teacher preparation: Network leadership, organization, and operation. *Journal of Teacher Education, 66*(5), 482–493. DOI: 10.1177/0022487115602312

MATHCOUNTS. (n.d.). https://www.mathcounts.org/

Mathematical Association of America (MAA). (n.d.). *American mathematics competitions*. Retrieved from http://www.maa.org/math-competitions

Mathematical Association of America (MAA). (n.d.). USA Mathematical Olympiad. Retrieved from http://www.maa.org/news/usa-mathematical-olympiad

Mathematical Association of America (MAA). (2015/2016). Instructional practice guide planned. *MAA Focus, 35*(6), 8.

Mathematics Teacher Education Partnership (MTE Partnership). (2014). *Guiding principles for secondary mathematics teacher preparation.* Washington, DC: Association of Public and Land-grant Universities.

McCulloch, A. W., Marshall, P. L., DeCuir-Gunby, J. T., & Caldwell, T. S. (2013). Math autobiographies: A window into teachers' identities as mathematics learners. *School Science and Mathematics, 113*(8), 380–389.

Meaney, T., Trinnick, T., & Fairhall, U. (2013). One size does not fit all: Achieving equity in Maori mathematics classrooms. *Journal for Research in Mathematics Education, 44*(1), 235–263.

Mistretta, R. M. (2013). "We do care," say parents. *Teaching Children Mathematics, 19*, 572–580.

Moll, L. C., Amanti, C., Neff, D., & Gonzalez, N. (1992). Funds of knowledge for teaching: Using a qualitative approach to connect homes and classrooms. *Qualitative Issues in Educational Research, 31*(2), 132–141.

Morin, J., & Samelson, V. M. (2015). Count on it: Congruent manipulative displays. *Teaching Children Mathematics, 21*, 363–370.

Moschkovich, J. (1999). Supporting the participation of English language learners in mathematical discussions. *For the learning of mathematics, 19*(1), 11–19.

Moschkovich, J. (2002). A situated and sociocultural perspective on bilingual mathematics learners. *Mathematical Thinking and Learning, 4*(2–3), 189–212.

Moschkovich, J. (2010). Language and mathematics education: Multiple perspectives and directions for research. Charlotte, NC: Information Age.

Moschkovich, J. (2012, April). Mathematics, the Common Core, and language: Recommendations for mathematics instruction for ELs aligned with the Common Core. In K. Hakuta & M. Santos (Eds.), *Understanding language: Language, literacy, and learning in the content areas* (pp. 17–31). Commissioned papers on language and literacy issues in the Common Core State Standards and Next Generation Science Standards. Stanford, CA: Stanford University Press.

Moser, J. M., & Carpenter, T. P. (1982). Young children are good problem solvers. *Arithmetic Teacher, 30*(3), 24–26.

Moyer-Packenham, P., Niezgoda, D., & Stanley, J. (2005). Young children's use of virtual manipulatives and other forms of mathematical representations. In W. J. Masalski & P. C. Elliott (Eds.), *Technology-supported mathematics learning environments: Sixty-seventh yearbook* (pp. 17–34). Reston, VA: National Council of Teachers of Mathematics.

National Association for Professional Development Schools. (2008). *What it means to be a professional development school.* Retrieved from http://napds.org/wp-content/uploads/2014/10/Nine-Essentials.pdf

National Association for the Education of Young Children (NAEYC). (2010). *NAEYC standards for early childhood professional preparation.* Washington, DC: Author.

National Association for the Education of Young Children (NAEYC). (2012). *2010 NAEYC standards for initial & advanced early childhood professional preparation programs.* Retrieved from http://www.naeyc.org/caep/files/caep/NAEYC%20Initial%20and%20Advanced%20Standards%2010_2012.pdf

National Board for Professional Teaching Standards (NBPTS). (2010). *National Board for Professional Teaching Standards mathematics standards* (3rd ed.). Washington, DC: U.S. Department of Education.

National Board for Professional Teaching Standards (NBPTS). (2012a). *National Board for Professional Teaching Standards early childhood generalist standards* (3rd ed). Washington, D.C.: U.S. Department of Education.

National Board for Professional Teaching Standards (NBPTS) (2012b). *National Board for Professional Teaching Standards middle childhood generalist standards* (3rd ed.). Washington, DC: U.S. Department of Education.

National Center for Education Statistics. (2015). *Children and youth with disabilities.* Retrieved from http://nces.ed.gov/programs/coe/pdf/coe_cgg.pdf

National Council of Teachers of Mathematics. (1991). *Professional standards for teaching mathematics.* Reston, VA: Author.

National Council of Teachers of Mathematics. (1995). *Assessment standards for school mathematics*. Reston, VA: Author.

National Council of Teachers of Mathematics. (2000). *Principles and standards for school mathematics.* Reston, VA: Author.

National Council of Teachers of Mathematics. (2009). *Focus in high school mathematics: Reasoning and sense making.* Reston, VA: Author.

National Council of Teachers of Mathematics. (2012). *NCTM CAEP mathematics content for secondary: Addendum to the NCTM CAEP Standards 2012.* Reston, VA: Author.

National Council of Teachers of Mathematics. (2014a). *Principles to actions: Ensuring mathematical success for all*. Reston, VA: Author.

National Council of Teachers of Mathematics. (2014b). *Access and equity in mathematics education: A position of the National Council of Teachers of Mathematics*. Reston, VA: Author.

National Council of Teachers of Mathematics. (2016a). *Computer science and mathematics education: A position of the National Council of Teachers of Mathematics*. Downloaded from http://www.nctm.org/uploadedFiles/Standards_and_Positions/Position_Statements/Computer%20science%20and%20math%20ed%20022416.pdf

National Council of Teachers of Mathematics. (2016b). *The NCTM High School Mathematics Task Force.* Retrieved from https://www.nctm.org/About/President,-Board-and-Committees/Committees-and-Appointees/High-School-Mathematics-Task-Force/

National Council of Teachers of Mathematics and the Council for the Accreditation of Educator Preparation (NCTM & CAEP). (2012a). *NCTM CAEP Standards—Middle grades (Initial preparation)*. Retrieved from http://www.nctm.org/uploadedFiles/Standards_and_Positions/CAEP_Standards/NCTM%20CAEP%20Standards%202012%20-%20Middle%20Grades.pdf

National Council of Teachers of Mathematics and the Council for the Accreditation of Educator Preparation (NCTM & CAEP). (2012b). *NCTM CAEP Standards— Secondary (Initial Preparation)*. Retrieved from http://www.nctm.org/uploadedFiles/Standards_and_Positions/CAEP_Standards/NCTM%20CAEP%20Standards%202012%20-%20Secondary.pdf

National Governors Association Center for Best Practices & Council of Chief State School Officers (NGA & CCSSO). (2010). *Common Core State Standards for Mathematics.* Washington, DC: Authors.

National Junior Mathematics Club. (n.d.). http://www.chialphamu.org/

National Mathematics Advisory Panel. (2008). *Foundations for success: The final report of the National Mathematics Advisory Panel*. Washington DC: U.S. Department of Education, Office of Planning, Evaluation and Policy Development.

National Research Council (NRC). (2001a). *Adding it up: Helping children learn mathematics*. J. Kilpatrick, J. Swafford, & B. Findell, (Eds.). Mathematics Learning Study Committee, Center for Education, Division of Behavioral and Social Sciences and Education. Washington, DC: National Academy Press.

National Research Council (NRC). (2001b). *Eager to learn: Educating our preschoolers*. Washington, DC: National Academies Press.

National Research Council NRC). (2007). *Taking science to school: Learning and teaching sciences in Grades K–8*. Washington, DC: National Academies Press.

National Research Council (NRC). (2009). *Mathematics learning in early childhood: Paths toward excellence and equity*. Washington, DC: National Academies Press.

National Research Council (NRC). (2010). The Teacher Development Continuum in the United States and China: Summary of a workshop. Washington, DC: National Academies Press.

National Research Council (NRC). (2012). *Engineering in K–12 education: Understanding the status and improving the prospects* (L. Katehi, G. Pearson, & M. Feder, Eds.). Washington, DC: National Academies Press.

National Research Council. (2013). *The mathematical sciences in 2025*. Washington, D.C.: National Academies Press.

Oakes, J. (2005). *Keeping track: How schools structure inequality*. New Haven, CT: Yale University Press.

Oakes, J. (2008). Keeping track: Structuring equality and inequality in an era of accountability. *Teachers College Record*, *110*, 700–712.

Odyssey of the Mind. (n.d.). Retrieved from http:// odysseyofthemind.com/

Olson, T. A., Olson, M., & Slovin, H. (2015). Putting essential understandings of ratios and proportions into practice in Grades 6–8 (H. Slovin, Ed.). In B. J. Dougherty (Series Ed.), *Putting essential understanding into practice series.* Reston, VA: National Council of Teachers of Mathematics.

Otto, A. D., Caldwell, J., Lubinski, C. A., & Hancock, S. W. (2011). Developing essential understanding of multiplication and division for teaching mathematics in Grades3–5 (E. C. Rathmell, Ed.). In R. M. Zbiek (Series Ed.), *Essential understanding series.* Reston, VA: National Council of Teachers of Mathematics.

Peck, R., Gould, R., & Miller, S.(2013). Developing essential understanding of statistics in Grades 9–12. In R. M. Zbiek (Series Ed.), *Essential understanding series.* Reston, VA: National Council of Teachers of Mathematics.

Peressini, D. (1998). The portrayal of parents in the school mathematics reform literature: Locating the context for parental involvement. *Journal for Research in Mathematics Education, 29*(5), 555–582.

Perkins, I., & Flores, A. (2002). Mathematical notations and procedures of recent immigrant students. *Mathematics Teaching in the Middle School*. 7(6) 346–351.

Peters, S. A., McGatha, M., Bush, W. S., Kezdy, A., Jones, J., Sato, T., Nussbaum, S., & Davis, R. (2016a). Diagnostic Teacher Assessments in Mathematics and Science (DTAMS): Middle grades geometry (Forms 1-4) [Assessment Instrument]. Louisville, KY: University of Louisville.

Peters, S. A., McGatha, M., Bush, W. S., Kezdy, A., Jones, J., Sato, T., Nussbaum, S., & Davis, R. (2016b). Diagnostic Teacher Assessments in Mathematics and Science (DTAMS): Middle grades statistics and probability (Forms 1-4) [Assessment Instrument]. Louisville, KY: University of Louisville.

Philipp, R. A. (2007). Mathematics teachers' beliefs and affect. In F. K. Lester, Jr. (Ed.), *Second handbook of research on mathematics teaching and learning* (pp. 257–315). Charlotte, NC: Information Age.

Philipp, R. A., Ambrose, R., Lamb, L. L. C., Sowder, J. T., Schappelle, B. P., Sowder, L., & Chauvot, J. (2007). Effects of early field experiences on the mathematical content knowledge and beliefs of prospective elementary school teachers: An experimental study. *Journal for Research in Mathematics Education, 38*, 438–476.

Ranta, J., & Dickey, E. (2015). *Secondary mathematics teacher recruitment campaign implementation guide.* Washington, DC: Association of Public Land-grant Universities. Retrieved from http://bit.ly/MATHImplGuide

Rigelman, N. M., & Ruben, B. (2012). Creating foundations for collaboration in schools: Utilizing professional learning communities to support teacher candidate learning and visions of teaching. *Teaching and Teacher Education, 28* (7), 979–989. DOI: http://dx.doi.org/10.1016/j.tate.2012.05.004.

Rodriguez, A. J., & Kitchen, R. S. (2005). *Preparing mathematics and science teachers for diverse classrooms: Promising strategies for transformative pedagogy.* Mahwah, NJ: Erlbaum.

Ronau, R. N., Peters, S. A., Bush, W. S., Maron, M., Niemi, R., Broyles, J., Hepp, D., & McGatha, M. (2016a). Diagnostic Teacher Assessments in Mathematics and Science (DTAMS): Middle grades algebraic thinking (Forms 1–4) [Assessment Instrument]. Louisville, KY: University of Louisville.

Ronau, R. N., Peters, S. A., Bush, W. S., Maron, M., Niemi, R., Broyles, J., Hepp, D., & McGatha, M. (2016b). Diagnostic Teacher Assessments in Mathematics and Science (DTAMS): Middle grades number and computation (Forms 1-4) [Assessment Instrument]. Louisville, KY: University of Louisville.

Roth McDuffie, A., Foote, M. Q., Bolson, C., Turner, E. E., Aguirre, J. M., Bartell, T. G., Drake, C., & Land, T. (2014). Using video analysis to support prospective K–8 teachers' noticing of students' multiple mathematical knowledge bases. *Journal of Mathematics Teacher Education*, 17, 245–270.

Rousseau, C., & Tate, W. F. (2003). No time like the present: Reflecting on equity in school mathematics. *Theory into Practice, 42*(3), 210–216.

Rudd, L. C., Lambert, M. C., Satterwhite, M., & Zaier, A. (2008). Mathematical language in early childhood settings: What really counts? *Early Childhood Education Journal, 36*(1), 75–80.

Sarama, J., & Clements, D. H. (2009). *Early childhood mathematics education research: Learning trajectories for young children.* New York, NY: Routledge.

Saul, J., Wandersman, A., Flaspohler, P., Duffy, J., Lubell, K., & Noonan, R. (2008). Research and action for bridging science and practice in prevention. *American Journal of Community Psychology, 41*(3–4), 165–170. doi:10.1007/s10464-008-9169-9

Schielack, J., Charles, R., Clements, D., Duckett, P., Fennell, F., Lewandowski, S., Trevino, E., & Zbiek, R. M. (2006). *Curriculum focal points for prekindergarten through Grade 8 mathematics: A quest for coherence.* Reston, VA: National Council of Teachers of Mathematics.

Schoenfeld, A. H. (2010). *How we think: A theory of goal oriented decision-making and its educational applications*. New York, NY: Routledge.

Schuck, S., & Russell, T. (2005). Self-study, critical friendship, and the complexities of teacher education. *Studying Teacher Education, 1*(2), 107–121.

Seymour, E., & Hewitt, N. M. (1997). *Talking about leaving: Why undergraduates leave the sciences*. Boulder, CO: Westview Press.

Shaughnessy, J. M., Chance, B., & Kranendonk, H. (2009). *Focus in high school mathematics: Reasoning and sense making in probability and statistics.* Reston, VA: National Council of Teachers of Mathematics.

Shulman, L. S. (1986). Those who understand: Knowledge growth in teaching. *Educational Researcher*, 15(2), 4–14.

Simic-Muller, K. (2015). Social justice and proportional reasoning. *Mathematics Teaching in the Middle School, 21*(3), 162–168.

Sinclair, N., Pimm, D., & Skelin, M. (2012a). Developing essential understanding of geometry for teaching mathematics in Grades 6–8. In R. M. Zbiek (Series Ed.), *Essential understanding series.* Reston, VA: National Council of Teachers of Mathematics.

Sinclair, N., Pimm, D., & Skelin, M. (2012b). Developing essential understanding of geometry for teaching mathematics in Grades 9–12. In R. M. Zbiek (Series Ed.), *Essential understanding series.* Reston, VA: National Council of Teachers of Mathematics.

Skemp, R. R. (1976). Relational understanding and instrumental understanding. *Mathematics Teaching, 77*, 20–26.

Smith, M. S. (2001). Practice-based professional development for teachers of mathematics. Reston, VA: National Council of Teachers of Mathematics.

Smith, M. S., Bill, V., & Hughes, E. K. (2008). Successfully implementing high-level tasks. *Mathematics Teaching in the Middle School, 14*(132–138).

Sowder, J. (2007). The mathematical education and development of teachers. In F. K. Lester (Ed.), Second handbook of research on mathematics teaching and learning (157–224). Reston, VA: National Council of Teachers of Mathematics.

Staub, F. C., & Stern, E. (2002). The nature of teachers' pedagogical content beliefs matters for students' achievement gains: Quasi-experimental evidence from elementary mathematics. *Journal of Educational Psychology, 94*(2), 344–355.

Steele, M. D., & Hillen, A. F. (2012). The content-focused methods course: A model for integrating pedagogy and mathematics content. *Mathematics Teacher Educator, 1*(1), 53–70.

Stein, M. K., Engle, R. A., Smith, M. S., & Hughes, E. K., (2008). Orchestrating productive mathematical discussions: Five practices for helping teachers move beyond show and tell. *Mathematical Thinking and Learning, 10*, 313–340.

Stein, M. K., Grover, B. W., & Henningsen, M. (1996). Building student capacity for mathematical thinking and reasoning: An analysis of mathematical tasks used in reform classrooms. *American Educational Research Journal, 33*(2), 455–488.

Stein, M. K., Smith, M. S., Henningsen, M. A., & Silver, E. A. (2000). *Implementing standards-based mathematics instruction: A casebook for professional development*. Reston, VA: National Council of Teachers of Mathematics.

Stein, M. K., Smith, M. S., Henningsen, M. A., & Silver, E. A. (2009). *Implementing standards-based mathematics instruction: A casebook for professional development* (2nd ed.). New York: Teachers College Press.

Steinthorsdottir, O. B. (n.d.). *Activities for developing mathematical practices in methods courses.* Unpublished paper. Ames: University of Northern Iowa.

Stenmark, J., Thompson, V., & Cossey, R. (1986). *Family math*. Berkeley, CA: Lawrence Hall of Science, Regents, University of California.

Stiff, L. V., & Johnson, J. L. (2011). Mathematical reasoning and sense-making begins with the opportunity to learn. In M. E. Strutchens & J. R. Quander (Eds.), *Focus in high school mathematics: Fostering reasoning and sense making for all students* (pp. 85–100). Reston, VA: National Council of Teachers of Mathematics.

Stigler, J. W., & Hiebert, J. (1999). The teaching gap: Best ideas from the world's teachers for improving education in the classroom. New York, NY: The Free Press.

Stigler, J. W., & Hiebert, J. (2004). Improving mathematics teaching. *Educational Leadership, 61*(5), 12–16.

Strutchens, M. E. (2002). Multicultural literature as a context for problem solving: Children and parents learning together. *Teaching Children Mathematics, 8*(8), 448–455.

Strutchens, M. E. (2012, June). *The need to change in secondary mathematics teacher preparation*. Presentation to the 5th annual MTE-Partnership Conference, Atlanta, GA.

Strutchens, M., Bay-Williams, J., Civil, M., Chval, K., Malloy, C. E., White, D. Y., D'Ambrosio, B., & Berry, R. Q. (2012). Foregrounding equity in mathematics teacher education. *Journal of Mathematics Teacher Education, 15*(1), 1–7.

Supovitz, J., Ebby, C. B., & Sirinides, P. (2013). TASK. Teacher analysis of student knowledge: A measure of learning trajectory-oriented formative assessment. Retrieved from Consortium for Policy Research in Education (CPRE) website: http://cpre.org/sites/default/files/researchreport/1446_taskreport.pdf

Swars, S. L., Smith, S. Z., Smith, M. E., & Hart, L. C. (2009). A longitudinal study of effects of a developmental teacher preparation program on elementary prospective teachers' mathematics beliefs. *Journal of Mathematics Teacher Education*, *12*(1), 47–66.

Sztajn, P., Confrey, J., Wilson, P. H., & Edgington, C. (2012). Learning trajectory based instruction: Toward a theory of teaching. *Educational Researcher*, *41*(5), 147–156.

Tallman, M., Carlson, M. P., Bressoud, D., & Pearson, M. (2016). A characterization of calculus I final exams in U.S. colleges and universities. *International Journal of Research in Undergraduate Mathematics Education*, *2*(1), 105–133.

Tate, W. F. (2005). *Access and opportunities to learn are not accidents: Engineering mathematical progress in your school*. Greensboro: Southeast Eisenhower Regional Consortium for Mathematics and Science Education, University of North Carolina at Greensboro.

Tatto, M.T., & Senk, S. (2011). The mathematics education of future primary and secondary teachers: Methods and findings from the Teacher Education and Development Study in Mathematics. *Journal of Teacher Education*, 62, 121–137.

Taylor, C. E., Rehm, M. A., & Catepillán, X. (2015). Maya calendars in the classroom. *Mathematics Teaching in the Middle School, 21*(2), 106–113.

Tsamir, P., & Tirosh, D. (2009). Affect, subject matter knowledge, and pedagogical content knowledge: The case of a kindergarten teacher. In J. Maasz & W. Schlöglmann (Eds.), *Beliefs and attitudes in mathematics education: New research results* (pp. 19–32). Rotterdam, the Netherlands: Sense.

Tucker, A., Burroughs, E., & Hodge, A. (2015). *A professional program for preparing future high school mathematics teachers*. In M. Siegel, C. Schumacher & P. Zorn (Eds.), 2015 CUPM curriculum guide to majors in the mathematical sciences. Washington, D.C.: Mathematical Association of America. Retrieved from http://www.maa.org/sites/default/files/HighSchoolMathematicsTeachersPASGReport.pdf

Turner, E. E., Celedón-Pattichis, S., Marshall, M., & Tennison, A. (2009). "Fijense amorcitos, les voy a contra una historia": The power of story to support solving and discussing mathematical problems among Latino and Latina kindergarten students. In D. Y. White & J. S. Spitzer (Eds.), *Responding to diversity: Grades pre-K–5* (pp. 23–42). Reston, VA: National Council of Teachers of Mathematics.

Turner, E. E., Drake, C., McDuffie, A. R., Aguirre, J., Bartell, T. G., & Foote, M. Q. (2012). Promoting equity in mathematics teacher preparation: A framework for advancing teacher learning of children's multiple mathematics knowledge bases. *Journal of Mathematics Teacher Education, 15*(1), 67–82.

U.S. Department of Education. (2013). National Center for Education Statistics, Schools and Staffing Survey (SASS), "Public School Teacher Data File" 1987-88 through 2011-12; "Private School Teacher Data File." Retrieved from https://nces.ed.gov/programs/digest/d14/tables/dt14_209.10.asp

Usiskin, Z. (2004). The importance of the transition years. Retrieved from http://ucsmp.uchicago.edu/resources/conferences/2004-05-01/

Usiskin, Z. (2015). Mathematical modeling. *Mathematics Teaching in the Middle School, 20*(8), 476–482.

Valencia, R. R. (2010). *Dismantling contemporary deficit thinking: Educational thought and practice*. New York, NY: Routledge.

Valero, O. (2009). What has power got to do with mathematics education? In P. Ernest, B. Greer, & B. Sriraman (Eds.), *Critical issues in mathematics education* (pp. 237–254). Charlotte, NC: Information Age.

Van Oers, B. (2010). Emergent mathematical thinking in the context of play. *Educational Studies in Mathematics, 74*(1), 23–37.

Vomvoridi-Ivanović, E. (2012). Using culture as a resource in mathematics: The case of four Mexican American prospective teachers in a bilingual after-school program. *Journal of Mathematics Teacher Education, 15*, 53–66.

Vomvoridi-Ivanovic, E., & Chval, K. B. (2014). Challenging beliefs and developing knowledge in relation to teaching English language learners: Examples from mathematics teacher education. In B. Cruz, C. Ellerbrock, A. Vasquez, & E. Howes (Eds.), *Talking diversity with teachers and teacher educators: Exercises and critical conversations across the curriculum* (pp. 115–130). New York, NY: Teachers College Press.

Wager, A. A. (2012). Incorporating out-of-school mathematics: From cultural context to embedded practice. *Journal of Mathematics Teacher Education, 15*(1), 9–23.

Wagner, D., & Lunney Borden, L. (2015). Common sense and necessity in (ethno)mathematics. In K. S. Sahin & R. S. Turner (Eds.), *New ground: Pushing the boundaries of studying informal learning in science, mathematics, and technology* (pp. 113–127). Rotterdam, The Netherlands: Sense. DOI 10.1007/978-94-6300-022-2_7

Weaver, D., Dick, T., & Rigelman, N. M. (2005, September). *Assessing the quantity and quality of student discourse in mathematics classrooms.* Paper presented at Math Science Partnership Evaluation Summit, Minneapolis, MN. Retrieved from http://hub.mspnet.org/entry.cfm/12626

Webb, D. (2016). Applying principles for active learning to promote student engagement in undergraduate calculus. *Proceedings of the 13th International Congress of Mathematics Education (ICME).* Hamburg, Germany: ICME.

White, A. L., Perry, B., Way, J., & Southwell, B. (2006). Mathematical attitudes, beliefs and achievement in primary preservice mathematics teacher education. *Mathematics Teacher Education and Development*, 7, 33–52.

Whitin, P., & Whitin, D. J. (2000). *Math is language too: Talking and writing in the mathematics classroom.* Urbana, IL: National Council of Teachers of English.

Wiliam, D., & Leahy, S. (2015). *Embedding formative assessment: Practical techniques for K–12 classrooms*. West Palm Beach, FL: Learning Sciences International.

Wilson, M., & Cooney, T. (2002). Mathematics teacher change and developments: The role of beliefs. In G. Leder, E. Pehkonen, & G. Torner (Eds.), *Beliefs: A hidden variable in mathematics education?* (pp. 127–147), Dordrecht, The Netherlands: Kluwer.

Wilson, S. M., Floden, R. E., & Ferrini-Mundy, J. (2001). *Teacher preparation research: Current knowledge, gaps, and recommendations* (Document R-01-3). University of Washington: Center for the Study of Teaching and Policy.

Yackel, E., & Cobb, P. (1996). Sociomathematical norms, argumentation, and autonomy in mathematics. *Journal for Research in Mathematics Education*, *27*, 458–477.

Zahner, W. (2015). The rise and run of a computational understanding of slope in a conceptually focused bilingual algebra class. *Educational Studies in Mathematics, 88*(1), 19–41. DOI: 10.1007/s10649-014-9575-x.

Zeichner, K., & Gore, M. (1990). *Teacher socialization.* In R. W. Huston (Ed.), *Handbook of research on teacher education* (pp. 329–348). New York, NY: Macmillan.

Printed and bound by CPI Group (UK) Ltd, Croydon, CR0 4YY
15/09/2026
14962341-0001